PRINCE PHAAHLE

The world of atlas of wisdom

Bridging East and west,North and South

First published by Tshireparadise 2024

Copyright © 2024 by Prince Phaahle

All rights reserved. No part of this publication may be reproduced, stored or transmitted in any form or by any means, electronic, mechanical, photocopying, recording, scanning, or otherwise without written permission from the publisher. It is illegal to copy this book, post it to a website, or distribute it by any other means without permission.

This novel is entirely a work of fiction. The names, characters and incidents portrayed in it are the work of the author's imagination. Any resemblance to actual persons, living or dead, events or localities is entirely coincidental.

Prince Phaahle asserts the moral right to be identified as the author of this work.

Prince Phaahle has no responsibility for the persistence or accuracy of URLs for external or third-party Internet Websites referred to in this publication and does not guarantee that any content on such Websites is, or will remain, accurate or appropriate.

Designations used by companies to distinguish their products are often claimed as trademarks. All brand names and product names used in this book and on its cover are trade names, service marks, trademarks and registered trademarks of their respective owners. The publishers and the book are not associated with any product or vendor mentioned in this book. None of the companies referenced within the book have endorsed the book.

First edition

Advisor: Tshireparadise

*This book was professionally typeset on Reedsy.
Find out more at reedsy.com*

Contents

1	Introduction	1
2	Chapter 1:The roots of wisdom	4
3	Chapter 2:The terrain of the mind	8
4	Chapter 3:The compass of the heart	12
5	Chapter 4:The horizon of purpose	16
6	Chapter 5:The ocean of transformation	20
7	Chapter 6:The mountain of mastery	25
8	Chapter 7:The river of renewal	31
9	Chapter 8:The ocean of oneness	35
10	Chapter 9:The temple of timelessness	39
11	Chapter 10:The garden of enlightenment	44
12	Chapter 11:The Oasis on inner peace	52
13	Chapter 12:The summit of enlightenment	58
14	Chapter 13:The sanctuary of legacy	65
15	Chapter 14:The horizon of possibility	71
16	Chapter 15:The haven of higher understanding	77
17	Chapter 16:The refuge of reflection	81
18	Chapter 17:The sanctuary of gratitude	84
19	Chapter 18:The horizon of Eternal wisdom	88
20	Epilogue	92
21	Glossary	94
22	Bibliography	97
23	Acknowledgment	99
24	Final thoughts	101

25	Conclusion	103
26	Endnotes	105
27	Author information	107

1

Introduction

Welcome to "The World Atlas of Wisdom," a captivating journey through the realms of knowledge, insight, and understanding. This comprehensive guide charts a course through the vast expanse of human experience, gathering the collective wisdom of sages, thinkers, and visionaries from across cultures and centuries.

Within these pages, we embark on an extraordinary quest to distill the essence of wisdom, exploring its many facets, forms, and expressions. From the timeless teachings of ancient philosophers to the cutting-edge discoveries of modern thought leaders, this atlas navigates the complex landscape of human wisdom.

Through a rich tapestry of stories, anecdotes, and reflections, "The World Atlas of Wisdom" invites you to:

- Explore the contours of wisdom, from its roots in ancient

traditions to its modern applications

- Discover the insights of luminaries, from Confucius to Einstein, and from Rumi to Maya Angelou

- Engage with the fundamental questions of existence, purpose, and meaning

- Develop a deeper understanding of yourself, others, and the world around you

Join us on this odyssey of discovery, as we map the uncharted territories of the human experience. "The World Atlas of Wisdom" is your passport to a world of profound insights, timeless truths, and transformative ideas.

About This Book

"The World Atlas of Wisdom" is organized into chapters, each focusing on a distinct aspect of wisdom, including:

["The Roots of Wisdom," "The Power of Intuition," "The Art of Resilience"]

Throughout this journey, you'll encounter:

- Inspirational quotes and passages from visionary thinkers

- Engaging stories and anecdotes from diverse cultures and traditions

- Reflective exercises and prompts to deepen your understanding

- Beautiful illustrations and visualizations to illuminate key concepts

INTRODUCTION

Embark on this transformative journey, and unlock the treasures of wisdom.

2

Chapter 1:The roots of wisdom

Wisdom's ancient roots stretch deep into the fertile soil of human experience, nourished by the wisdom of countless generations. From the dawn of civilization, wise men and women have sought to understand the mysteries of existence, sharing their insights through timeless teachings, sacred texts, and oral traditions. The whispers of wisdom from ancient cultures – Egypt, Greece, China, and India – still resonate within us, offering guidance for navigating life's complexities. In the words of the Greek philosopher, Heraclitus, "The way up and the way down are one and the same." This paradoxical truth hints at wisdom's capacity to reconcile contradictions, revealing the interconnectedness of all things.

As we explore wisdom's roots, we find ourselves in the company of visionary thinkers who dared to question the status quo. Confucius, Lao Tzu, and Buddha – giants of Eastern wisdom – taught us to cultivate self-awareness, compassion, and inner balance. Meanwhile, Socrates, Plato, and Aristotle – pillars of Western philosophy – probed the nature of reality, ethics, and human potential. Their collective wisdom reminds us that true understanding requires embracing paradox, ambiguity, and uncertainty. The ancient Greeks' concept of "phronesis" – practical wisdom – reminds us that wisdom is not merely theoretical, but a lived experience.

The wisdom of indigenous cultures and traditional societies offers another vital thread in the tapestry of human understanding. The Aboriginal Dreamtime, African diasporic traditions, and Native American storytelling all reveal the intricate web of relationships between humans, nature, and the cosmos. These wisdom traditions emphasize the importance of community, reciprocity, and reverence for the land and its inhabitants. By listening to the elders and honoring the past, we may rediscover the timeless wisdom hidden within our own cultural heritage.

As we venture deeper into the world atlas of wisdom, we'll encounter numerous exemplars of wise living – from spiritual leaders and philosophers to artists, scientists, and social visionaries. Their stories and teachings will illuminate the many paths to wisdom, demonstrating that this journey is uniquely personal, yet universally shared. Whether drawn from ancient texts, contemporary insights, or everyday experiences, wisdom's roots remind us that true understanding requires patience, humility, and an open heart.

As we delve into the roots of wisdom, we begin to appreciate the subtle interplay between tradition and innovation. Wisdom's evolution is a testament to human resilience and adaptability, as each generation builds upon the discoveries of the past while forging new paths. The timeless wisdom of the ancients harmonizes with modern insights, yielding a rich tapestry of knowledge and experience. From the Stoic philosophers' emphasis on reason and self-control to the contemplative practices of Buddhist mindfulness, wisdom's diverse threads weave together to form a majestic fabric of understanding.

The landscape of wisdom is dotted with iconic figures whose lives embodied the pursuit of insight and knowledge. Sages like Epicurus, who cultivated gardens of wisdom in ancient Athens, and visionaries like Rumi, who illuminated the intersection of love and spirituality, inspire us to tend our own gardens of wisdom. Their legacies remind us that wisdom is not solely the domain of the intellect, but a holistic embrace of heart, mind, and spirit. As we navigate the complexities of our own lives, wisdom's roots offer a sturdy foundation, grounding us in the timeless truths that transcend fleeting fads and fashionable ideologies.

The geography of wisdom stretches across vast expanses of time and space, connecting disparate cultures and civilizations. From the pyramids of Egypt to the temples of Angkor Wat, from the libraries of Alexandria to the scriptoriums of medieval Europe, wisdom's sacred sites beckon us to pilgrimage. Each location holds secrets, stories, and insights waiting to be unearthed, inviting us to participate in the ongoing conversation of human wisdom. As we embark on this journey, we become part of a global community bound together by a shared quest for understanding, compassion, and wisdom.

As we explore the cartography of wisdom, we begin to notice the convergence of seemingly disparate paths. The mystical and the rational, the spiritual and the scientific, the ancient and the modern – all find common ground in the pursuit of wisdom. This integration is reflected in the lives of polymaths like Leonardo da Vinci, who embodied the Renaissance ideal of unity between art and science. Similarly, wisdom's contemporary champions – thinkers like Integral philosopher Ken Wilber and systems theorist Fritjof Capra – demonstrate the power of synthesis, weaving together insights from multiple disciplines to illuminate our shared human experience.

3

Chapter 2:The terrain of the mind

The mind is wisdom's vast and mysterious terrain, encompassing the landscapes of thought, emotion, and perception. To navigate this inner world, we must first understand the geography of our own consciousness. The ancient Greeks distinguished between two modes of thinking: logos, the rational intellect, and mythos, the intuitive imagination. This dichotomy still resonates today, as we struggle to balance reason and intuition, analysis and creativity. Wisdom's journey requires embracing both aspects, integrating the logical and the imaginative to reveal the hidden patterns and connections that underlie reality.

As we venture deeper into the mind's terrain, we encounter the vast expanses of the subconscious, where emotions, memories, and desires reside. The wisdom traditions of the East – Buddhism, Taoism, and Advaita Vedanta – offer powerful tools for exploring this inner realm, from meditation and mindfulness to the subtle energies of chi and prana. These practices allow us to map the contours of our own psyche, uncovering hidden biases, fears, and motivations that shape our thoughts and actions.

CHAPTER 2: THE TERRAIN OF THE MIND

By illuminating the dark recesses of our own minds, we may discover the seeds of compassion, empathy, and understanding that will guide us on our journey.

The mind's terrain is also shaped by the forces of culture, history, and social conditioning. Our perceptions are filtered through the lens of language, upbringing, and collective narrative, influencing how we perceive ourselves, others, and the world. Wisdom demands that we question these assumptions, challenging our own biases and limitations to reveal the uncharted territories of the mind. This critical self-awareness is the hallmark of wisdom's seekers, from Socrates' examined life to the contemporary critiques of cultural narrative. By charting the complexities of our own mental terrain, we may uncover the hidden riches of wisdom, waiting to be unearthed.

As we explore the mind's vast expanse, we find ourselves at the threshold of the unknown, where the boundaries of self and other, thought and reality, begin to blur. Here, wisdom's ancient cartographers – mystics, shamans, and seers – have left behind cryptic maps and symbolic guides to navigate the unseen realms. From the Kabbalistic Tree of Life to the Tibetan Book of the Dead, these esoteric traditions offer keys to deciphering the mind's deepest mysteries. Will we dare to venture into these uncharted territories, where the very fabric of reality awaits our discovery?

As we venture deeper into the mind's terrain, we encounter the paradoxical landscape of duality, where opposing forces shape our perceptions and experiences. Light and darkness, order and chaos, certainty and uncertainty – these dichotomies weave the intricate tapestry of human existence. Wisdom's sages have long recognized the interdependence of these seeming opposites, revealing the hidden unity that underlies all contrasts. The Taoist principle of yin-yang, the Buddhist concept of interconnectedness, and the Hegelian dialectic all attest to the dynamic harmony that emerges from the tension between opposing forces.

The mind's terrain is also punctuated by the oases of insight, where sudden revelations illuminate the path ahead. These epiphanies often arise from the convergence of disparate threads – a chance encounter, a dream, or a long-forgotten memory. Wisdom's art lies in recognizing these synchronicities, weaving them into the narrative of our lives to reveal the hidden patterns and purposes that guide us. By attentive listening to the whispers of our own hearts, we may discern the subtle signals that navigate us through life's labyrinthine journey.

In this inner landscape, wisdom's travelers encounter the mirages of illusion, where the mind's projections and assumptions distort reality. The Buddhist notion of "maya" – the veil of ignorance – reminds us that our perceptions are filtered through the prism of our own conditioning. To traverse this treacherous terrain, we must cultivate the discerning eye of critical awareness, distinguishing between the mirage and the oasis, the illusion and the truth. By refining our perception, we may behold the world anew, unencumbered by the distortions of our own making.

As we navigate the mind's vast expanse, we come upon the crossroads of choice and destiny. Here, wisdom's ancient travelers – philosophers, mystics, and sages – offer divergent maps to guide our journey. Some propose the path of free will, where individual agency shapes the course of life. Others suggest the gentle currents of fate, where circumstances and karma guide us toward our highest purpose. Wisdom's synthesis reconciles these perspectives, revealing the dynamic interplay between choice and destiny. By acknowledging the forces beyond our control, we may exercise our freedom to choose, crafting a life of intention and purpose.

The mind's terrain is also home to the gardens of imagination, where creativity and innovation bloom. Wisdom's artistic visionaries – Shakespeare, Leonardo, and Picasso – demonstrate the transformative power of imagination, reshaping reality through the lens of their creative genius. By cultivating our own imaginative faculties, we may tap the boundless potential of the mind, conjuring solutions to life's challenges and manifesting new possibilities. As we tend the gardens of our imagination, wisdom's flowers of inspiration will blossom, guiding us toward unseen horizons.

In the mind's depths, wisdom's explorers encounter the mysterious realm of the collective unconscious, where archetypes and universal symbols reside. This shared reservoir of human experience – identified by Carl Jung and Joseph Campbell – holds the secrets of our shared humanity. By engaging with these timeless themes and motifs, we may access the wisdom of the ages, transcending the boundaries of individual experience. Through the collective unconscious, we connect with the greater whole, recognizing our place within the grand tapestry of human existence.

4

Chapter 3:The compass of the heart

The heart is wisdom's sacred compass, guiding us through life's complexities with compassion, empathy, and love. This inner navigator charts our course, balancing reason and emotion, intellect and intuition. Wisdom's sages have long recognized the heart's pivotal role, from the ancient Greek concept of "cardiac intelligence" to the Sufi mystics' emphasis on the heart's spiritual centrality. By listening to the heart's whispers, we may discern our deepest values, passions, and purposes.

As we explore the heart's terrain, we encounter the landscapes of emotion, where feelings and sentiments shape our experiences. Wisdom's emotional intelligence – exemplified by leaders like Nelson Mandela and Mahatma Gandhi – demonstrates the power of compassion, forgiveness, and empathy in transforming conflict and adversity. By cultivating emotional awareness, we may navigate life's turbulent waters with greater ease, depth, and resilience. The heart's emotional spectrum – from joy and gratitude to sorrow and grief – becomes a rich tapestry of wisdom, woven from the threads of our shared humanity.

CHAPTER 3: THE COMPASS OF THE HEART

The heart is also the seat of intuition, where instinct and instinctive knowing guide our decisions and actions. Wisdom's intuitive travelers – from indigenous shamans to contemporary visionaries – rely on the heart's subtle perceptions to navigate uncertainty and change. By trusting our intuition, we may access the hidden patterns and connections that underlie reality, uncovering innovative solutions and untapped potential. The heart's intuitive voice whispers truths beyond reason's reach, illuminating the unseen pathways to our highest good.

As we journey deeper into the heart's realm, we encounter the mysteries of love and connection. Wisdom's poets and mystics – Rumi, Hafiz, and Kabir – celebrate the transformative power of love, dissolving boundaries and uniting us in our shared humanity. By embracing love's wisdom, we may transcend the illusions of separation, recognizing our interconnectedness with all beings and the world around us. The heart's capacity for love becomes the ultimate compass, guiding us toward wisdom, compassion, and unity.

As we navigate the heart's vast expanse, we come upon the sacred sites of vulnerability and openness. Wisdom's courageous travelers – from Brené Brown to Desmond Tutu – demonstrate the transformative power of vulnerability, embracing uncertainty and risk to foster deeper connections and growth. By surrendering our defenses and embracing our susceptibility, we may discover the heart's hidden strengths: compassion, empathy, and unconditional love. Vulnerability becomes the gateway to wisdom's deepest treasures: intimacy, trust, and authentic relationship.

The heart's terrain is also home to the wisdom of forgiveness, where release and letting go free us from the burdens of resentment and pain. Wisdom's forgiveness is not forgetting or condoning, but rather releasing the emotional charge that binds us to past hurts. By forgiving ourselves and others, we may reclaim our energy, rediscover our compassion, and rebirth our sense of purpose. The heart's capacity for forgiveness becomes a balm to our souls, healing the wounds of yesterday to illuminate the path ahead.

In the heart's depths, wisdom's explorers encounter the mystical realm of unity and interconnectedness. This sacred space – described by mystics and sages across traditions – transcends the illusions of separation, revealing our essential oneness with all existence. By experiencing this unity, we may recognize the world as a vast, interconnected web of life, where every action ripples outward to touch the hearts of others. The heart's wisdom reminds us that our individual journeys are inextricably linked to the greater whole, guiding us toward compassion, cooperation, and collective evolution.

As we embark on the heart's journey, we encounter the mirrors of self-reflection, where introspection and self-awareness reveal our deepest truths. Wisdom's sages – from Marcus Aurelius to Maya Angelou – demonstrate the transformative power of self-reflection, inviting us to confront our shadows, biases, and limitations. By gazing into the heart's mirror, we may discern our authentic selves, releasing the masks and pretenses that conceal our true nature. Self-reflection becomes the heart's compass, navigating us through life's complexities and guiding us toward radical self-acceptance.

The heart's terrain is also punctuated by the oases of gratitude and appreciation, where joy and thankfulness nourish our souls. Wisdom's grateful travelers – from Thich Nhat Hanh to Nelson Mandela – show us how gratitude transmutes adversity into opportunity, transforming suffering into wisdom. By cultivating gratitude, we may reframe our perceptions, recognizing the hidden blessings and beauty that surround us. The heart's gratitude becomes a magnet, attracting abundance, compassion, and connection into our lives.

In the heart's depths, wisdom's explorers discover the hidden springs of resilience and hope. This inner wellspring – tapped by wisdom's resilient heroes, from Malala Yousafzai to Nelson Mandela – sustains us through life's darkest moments, illuminating the path forward. By embracing hope's wisdom, we may transcend despair, finding the courage to reimagine and rebuild. The heart's resilience becomes our guiding light, navigating us through uncertainty and guiding us toward a brighter future.

5

Chapter 4:The horizon of purpose

As we navigate the vast expanse of human experience, we arrive at the horizon of purpose, where our deepest longings and highest aspirations converge. Wisdom's sages have long recognized the importance of purpose in guiding our journey, from Aristotle's concept of "telos" to Viktor Frankl's logotherapy. By clarifying our purpose, we may align our actions, values, and passions, transforming our lives into a meaningful narrative. Purpose becomes the north star, illuminating our path through life's uncertainties and challenges.

The horizon of purpose stretches across the landscape of our lives, encompassing our relationships, work, and contributions to the world. Wisdom's purpose-driven travelers – from Jane Goodall to Martin Luther King Jr. – demonstrate the power of aligning our actions with our values, creating a ripple effect of positive change. By embracing our purpose, we may transcend the boundaries of self-interest, recognizing our interconnectedness with others and the world around us. Purpose becomes the bridge between our individuality and our shared humanity, guiding us toward a life of significance and impact.

As we approach the horizon of purpose, we encounter the mirrors of legacy and impact, reflecting our contributions to the world. Wisdom's legacy-minded leaders – from Nelson Mandela to Rachel Carson – show us how our actions can shape the future, inspiring generations to come. By considering our legacy, we may reframe our priorities, investing in the things that truly matter: love, compassion, and the well-being of all beings. The horizon of purpose becomes a call to action, inviting us to leave a lasting imprint on the world.

The journey to the horizon of purpose requires courage, resilience, and determination. Wisdom's wayfinders – from Maya Angelou to Nelson Mandela – demonstrate the power of perseverance, overcoming obstacles and setbacks to realize their vision. By embracing our purpose, we may discover our inner strength, navigating life's challenges with greater ease, confidence, and wisdom. The horizon of purpose becomes our guiding vision, illuminating the path to a life of meaning, purpose, and fulfillment.

As we gaze upon the horizon of purpose, we behold the majestic landscape of our highest potential. Wisdom's visionary leaders – from Buckminster Fuller to Wangari Maathai – demonstrate the transformative power of aligning our actions with our values, creating a better world for all. By embracing our purpose, we may unlock our innate creativity, innovation, and genius, unleashing a cascade of positive change. The horizon of purpose becomes a beacon, guiding us toward a future where our unique gifts and talents are fully expressed, benefiting humanity and the planet.

The journey to the horizon of purpose requires navigating the valleys of uncertainty and self-doubt. Wisdom's wayfinders – from Steve Jobs to Frida Kahlo – show us how to transform adversity into opportunity, finding purpose in the midst of challenge. By embracing our vulnerability, we may discover our inner resilience, cultivating the courage to pursue our passions and values. The horizon of purpose becomes a promise, reminding us that our lives have meaning, significance, and impact.

As we reach the horizon of purpose, we encounter the vast expanse of our collective potential. Wisdom's collaborative visionaries – from Martin Luther King Jr. to Malala Yousafzai – demonstrate the power of united action, mobilizing individuals and communities toward a common goal. By recognizing our interconnectedness, we may transcend the boundaries of individual achievement, co-creating a brighter future for all. The horizon of purpose becomes a celebration, honoring our diversity, creativity, and shared humanity.

As we embody our purpose, we become stewards of a larger legacy, transcending our individual lifetimes. Wisdom's timeless leaders – from Aristotle to Nelson Mandela – demonstrate the power of leaving a lasting impact, shaping the course of history. By integrating our values, passions, and talents, we may create a ripple effect of positive change, inspiring future generations. The horizon of purpose becomes a timeless inheritance, guiding humanity toward a brighter future.

CHAPTER 4: THE HORIZON OF PURPOSE

The journey to the horizon of purpose requires embracing our wholeness, integrating our shadows and light. Wisdom's holistic visionaries – from Carl Jung to Maya Angelou – show us how to reconcile our contradictions, becoming more authentic and whole. By acknowledging our complexities, we may tap into our inner wisdom, navigating life's paradoxes with greater ease and clarity. The horizon of purpose becomes a mirror, reflecting our truest selves.

As we approach the horizon of purpose, we encounter the threshold of transcendence, where our individuality merges with the universal. Wisdom's spiritual leaders – from Rumi to Einstein – demonstrate the power of connecting with the divine, recognizing our place within the grand tapestry of existence. By transcending our ego's boundaries, we may experience the oneness of all being, becoming instruments of love, compassion, and wisdom. The horizon of purpose becomes a gateway, leading us to the infinite possibilities of the human spirit.

6

Chapter 5:The ocean of transformation

As we navigate the vast expanse of human experience, we arrive at the ocean of transformation, where wisdom's deepest currents reshape our lives. This boundless sea – charted by sages, mystics, and visionaries – holds the power to rebirth our souls, revitalizing our purpose, passion, and potential. Transformation's waves crash upon the shores of our hearts, inviting us to surrender, let go, and rebirth.

The ocean of transformation stretches across the landscape of our lives, encompassing our darkest nights and brightest dawns. Wisdom's transformative travelers – from Buddha to Malcolm X – demonstrate the courage to confront their shadows, embracing the darkness to emerge radiant and renewed. By surrendering to transformation's tides, we may release our limitations, rediscover our essence, and rebirth our lives. The ocean's depths become a crucible, refining our character, compassion, and wisdom.

As we dive into the ocean's depths, we encounter the undertows

CHAPTER 5: THE OCEAN OF TRANSFORMATION

of our subconscious, where hidden fears, desires, and biases reside. Wisdom's psycho-spiritual explorers – from Jung to Assagioli – show us how to navigate these unseen realms, integrating our fragmented selves. By confronting our shadows, we may reclaim our wholeness, harnessing the transformative power of our psyche. The ocean's undertows become a catalyst, propelling us toward self-awareness, healing, and growth.

The ocean of transformation also holds the mysteries of spiritual awakening, where the boundaries of self dissolve into the infinite. Wisdom's mystical cartographers – from Rumi to Eckhart Tolle – map the terrain of enlightenment, guiding us toward the shores of unity. By embracing transformation's spiritual dimensions, we may transcend our ego's confines, experiencing the oneness of all existence. The ocean's horizon becomes a gateway, beckoning us to the timeless, spaceless realm of the divine.

As we navigate the ocean of transformation, we encounter the rip currents of change, where familiar shores disappear and new landscapes emerge. Wisdom's adaptive leaders – from Steve Jobs to Nelson Mandela – demonstrate the agility to ride the waves of uncertainty, reinventing themselves and their worlds. By embracing change's transformative power, we may release our attachment to the status quo, discovering innovative solutions and untapped potential. The ocean's currents become a dynamic force, propelling us toward growth, renewal, and evolution.

The ocean's depths also hold the treasures of forgiveness and release, where we surrender the burdens of resentment and pain. Wisdom's compassionate teachers – from Thich Nhat Hanh to Desmond Tutu – show us how to navigate the complex waters of forgiveness, freeing ourselves and others from the shackles of hurt. By releasing our emotional cargo, we may reclaim our energy, rediscover our compassion, and rebirth our relationships. The ocean's depths become a sanctuary, healing the wounds of yesterday to illuminate the path ahead.

As we explore the ocean's mysteries, we encounter the mirrors of self-reflection, where our true nature shines like a beacon. Wisdom's introspective sages – from Marcus Aurelius to Maya Angelou – demonstrate the power of self-awareness, illuminating the contours of our souls. By gazing into the ocean's mirror, we may discern our deepest values, passions, and purposes, aligning our lives with our essence. The ocean's reflections become a guiding light, navigating us toward authenticity, integrity, and wisdom.

As we dive deeper into the ocean of transformation, we discover the hidden treasures of resilience and adaptability. Wisdom's courageous explorers – from Ernest Shackleton to Malala Yousafzai – demonstrate the capacity to navigate uncertainty, finding strength in vulnerability. By embracing the ocean's turbulence, we may develop our inner compass, staying oriented amidst life's unpredictable currents. The ocean's depths become a crucible, forging our character, compassion, and wisdom.

The ocean's transformative power also awakens our creative potential, where imagination and innovation reshape our worlds. Wisdom's visionary artists – from Leonardo da Vinci to Frida Kahlo – show us how to tap the ocean's inspirational currents, channeling creativity into masterpieces of meaning. By surrendering to the ocean's muse, we may birth new possibilities, transforming our lives and the world around us. The ocean's waves become a symphony, orchestrating our unique contributions to humanity's grand narrative.

As we ride the ocean's waves, we encounter the threshold of transcendence, where individuality merges with the universal. Wisdom's spiritual leaders – from Rumi to Einstein – demonstrate the power of connecting with the divine, recognizing our place within the grand tapestry of existence. By transcending our ego's boundaries, we may experience the oneness of all being, becoming instruments of love, compassion, and wisdom. The ocean's horizon becomes a gateway, beckoning us to the infinite possibilities of the human spirit.

As we dive deeper into the ocean of transformation, we discover the hidden treasures of resilience and adaptability. Wisdom's courageous explorers – from Ernest Shackleton to Malala Yousafzai – demonstrate the capacity to navigate uncertainty, finding strength in vulnerability. By embracing the ocean's turbulence, we may develop our inner compass, staying oriented amidst life's unpredictable currents.

The ocean's transformative power also awakens our creative potential, where imagination and innovation reshape our worlds. Wisdom's visionary artists – from Leonardo da Vinci to Frida Kahlo – show us how to tap the ocean's inspirational currents, channeling creativity into masterpieces of meaning. By surrendering to the ocean's muse, we may birth new possibilities, transforming our lives and the world around us.

As we ride the ocean's waves, we encounter the threshold of transcendence, where individuality merges with the universal. Wisdom's spiritual leaders – from Rumi to Einstein – demonstrate the power of connecting with the divine, recognizing our place within the grand tapestry of existence. By transcending our ego's boundaries, we may experience the oneness of all being, becoming instruments of love, compassion, and wisdom.

In the ocean's depths, wisdom's ancient sages – from Hermes to Lao Tzu – reveal the secrets of the universe, hidden patterns and rhythms governing reality. By aligning ourselves with these timeless principles, we may harmonize our lives with the cosmos, discovering the hidden order beneath life's chaos. The ocean's depths become a sacred text, guiding us toward enlightenment and unity.

As we emerge from the ocean's transformative waters, we are reborn, renewed, and revitalized. Wisdom's resilient travelers – from Nelson Mandela to Jane Goodall – demonstrate the power of transformation, radiating hope and inspiration to a world in need. By embracing the ocean's wisdom, we may become beacons of light, illuminating the path for others and forging a brighter future for all humanity.

7

Chapter 6:The mountain of mastery

As we ascend the mountain of mastery, we enter the realm of expertise, where wisdom's disciplines converge. This lofty peak – climbed by sages, innovators, and visionaries – represents the pinnacle of human potential, where skill, knowledge, and experience unite. Mastery's majestic landscape stretches across the horizons of our lives, encompassing the arts, sciences, and humanities.

The mountain's base camp is anchored in the valleys of discipline, where focused effort and persistence lay the foundations for greatness. Wisdom's master craftsmen – from Michelangelo to Steve Jobs – demonstrate the power of dedication, honing their skills through relentless practice and experimentation. By embracing discipline's rigor, we may cultivate our inner strength, transforming raw talent into refined expertise.

As we ascend the mountain's slopes, we encounter the ridges of resilience, where adaptability and determination overcome obstacles. Wisdom's courageous innovators – from Marie Curie to Thomas Edison – show us how to navigate uncertainty, converting setbacks into stepping stones. By developing our resilience, we may transcend the boundaries of our potential, achieving breakthroughs that reshape the world.

The mountain's summit is shrouded in the mist of mystery, where wisdom's deepest secrets await discovery. Wisdom's visionary leaders – from Einstein to Martin Luther King Jr. – demonstrate the power of integrating knowledge, revealing the interconnectedness of all things. By embracing the mystery, we may transcend the fragmentation of knowledge, recognizing the unity that underlies human experience.

As we descend the mountain, we bring our mastery to the world, sharing our gifts with humanity. Wisdom's compassionate mentors – from Socrates to Maya Angelou – show us how to transmit our knowledge, inspiring others to climb the mountain. By sharing our mastery, we may leave a lasting legacy, illuminating the path for future generations.

The mountain of mastery stands as a testament to human potential, beckoning us to ascend to greatness. Wisdom's timeless sages remind us that mastery is not a destination, but a journey – a lifelong pursuit of excellence, fueled by curiosity, passion, and dedication.

The mountain's snow-capped peak glistens with the promise of transcendence, inviting us to surpass ourselves. Wisdom's spiritual leaders – from Buddha to Rumi – demonstrate the power of self-transcendence, merging individuality with the universal. By embracing the mountain's wisdom, we may

CHAPTER 6: THE MOUNTAIN OF MASTERY

become instruments of love, compassion, and wisdom, forging a brighter future for all humanity.

The mountain of mastery is a symbol of our collective potential, a beacon guiding us toward excellence. By embracing its wisdom, we may become the masters of our own destiny, crafting a world of beauty, truth, and wisdom.

As we explore the mountain's vast terrain, we discover the valleys of specialization, where expertise converges with passion. Wisdom's master artisans – from Stradivari to Georgia O'Keeffe – demonstrate the power of focused creativity, crafting works of timeless beauty. By embracing specialization's depth, we may refine our skills, innovating within the boundaries of our craft.

The mountain's rugged slopes also conceal the caves of introspection, where self-awareness illuminates our inner landscape. Wisdom's contemplative sages – from Marcus Aurelius to Thich Nhat Hanh – show us how to navigate the labyrinthine corridors of our minds, confronting fears, desires, and limitations. By exploring our inner depths, we may uncover our authentic selves, aligning our actions with our values.

As we near the mountain's summit, we encounter the ridges of integration, where disparate threads of knowledge converge. Wisdom's polymathic visionaries – from Leonardo da Vinci to Buckminster Fuller – demonstrate the power of synthesis, revealing the interconnectedness of all things. By integrating our knowledge, we may transcend disciplinary boundaries, unlocking innovative solutions to humanity's pressing challenges.

The mountain's majestic vistas offer panoramic views of our shared humanity, reminding us that mastery serves a greater purpose. Wisdom's compassionate leaders – from Nelson Mandela to Malala Yousafzai – show us how to harness our expertise for the greater good, uplifting others and transforming our world. By sharing our mastery, we may become agents of positive change, leaving a lasting legacy that inspires future generations.

Within the mountain's ancient stones lies the wisdom of the ages, whispering secrets to those who listen. Wisdom's timeless sages remind us that mastery is a lifelong journey, not a destination – a quest to refine our character, cultivate our compassion, and illuminate our understanding.

As we descend the mountain, we carry the wisdom of mastery back into our lives, applying its principles to forge a better world. Wisdom's practical innovators – from Steve Jobs to Jane Goodall – demonstrate the power of translating vision into action, turning ideas into reality. By embracing the mountain's wisdom, we may become the architects of our own destiny, crafting a future of beauty, truth, and wisdom.

As we explore the mountain's hidden valleys, we stumble upon the springs of creativity, where innovation and imagination converge. Wisdom's visionary artists – from Picasso to Frida Kahlo – demonstrate the power of creative expression, transforming the world through their unique perspectives. By tapping into our creative potential, we may bring forth new ideas, reshaping reality with every brushstroke.

CHAPTER 6: THE MOUNTAIN OF MASTERY

The mountain's rugged terrain also conceals the caverns of critical thinking, where discernment and analysis illuminate our understanding. Wisdom's philosophical sages – from Socrates to Immanuel Kant – show us how to navigate the labyrinthine corridors of reasoning, evaluating evidence and challenging assumptions. By honing our critical faculties, we may distinguish truth from falsehood, wisdom from folly.

As we ascend the mountain's upper reaches, we encounter the plateaus of perspective, where broad vistas reveal the interconnectedness of all things. Wisdom's holistic thinkers – from Buckminster Fuller to Fritjof Capra – demonstrate the power of systems thinking, recognizing the intricate web of relationships that binds our world. By embracing the bigger picture, we may transcend fragmented thinking, addressing the complex challenges of our time.

The mountain's windswept peaks remind us that mastery is a dynamic process, not a static state. Wisdom's adaptive leaders – from Alexander the Great to Nelson Mandela – show us how to navigate uncertainty, pivoting in response to changing circumstances. By embracing adaptability, we may stay relevant, effective, and wise in an ever-shifting world.

Within the mountain's ancient heart lies the wisdom of the ages, guiding us toward mastery's ultimate goal: self-realization. Wisdom's spiritual sages – from Buddha to Rumi – remind us that true mastery lies not in external achievements, but in cultivating our inner depths, aligning our actions with our values, and radiating love, compassion, and wisdom.

As we integrate the mountain's wisdom into our lives, we become the embodiment of mastery, inspiring others through our example. Wisdom's compassionate mentors – from Aristotle to Maya Angelou – demonstrate the power of mentorship, guiding others along the path to mastery. By sharing our wisdom, we may leave a lasting legacy, illuminating the journey for those who follow.

8

Chapter 7:The river of renewal

As we embark on the river of renewal, we enter the realm of transformation, where wisdom's currents reshape our lives. This majestic watercourse – navigated by sages, innovators, and visionaries – represents the eternal cycle of growth, decay, and rebirth. Renewal's soothing waters stretch across the landscape of our souls, nourishing our deepest longings for revitalization.

The river's source lies in the springs of self-awareness, where introspection illuminates our inner world. Wisdom's contemplative sages – from Marcus Aurelius to Thich Nhat Hanh – demonstrate the power of introspection, confronting fears, desires, and limitations. By exploring our inner depths, we may uncover our authentic selves, aligning our actions with our values.

As we flow downstream, we encounter the rapids of change, where uncertainty and risk reshape our lives. Wisdom's courageous innovators – from Steve Jobs to Malala Yousafzai – show us how to navigate turbulence, converting obstacles into opportunities. By embracing change's transformative power, we may release our attachment to the status quo, discovering innovative solutions to humanity's pressing challenges.

The river's tranquil stretches reveal the beauty of simplicity, where clarity and focus guide our journey. Wisdom's minimalist masters – from Henry David Thoreau to Marie Kondo – demonstrate the power of simplicity, stripping away unnecessary complexity. By embracing simplicity's elegance, we may rediscover our essential nature, aligning our lives with our deepest priorities.

As we approach the river's delta, we encounter the estuaries of integration, where disparate threads of knowledge converge. Wisdom's polymathic visionaries – from Leonardo da Vinci to Buckminster Fuller – show us how to synthesize diverse perspectives, revealing the interconnectedness of all things. By integrating our understanding, we may transcend disciplinary boundaries, unlocking innovative solutions to humanity's complex challenges.

The river's mouth opens into the vast ocean of transcendence, where individuality merges with the universal. Wisdom's spiritual leaders – from Rumi to Einstein – demonstrate the power of connecting with the divine, recognizing our place within the grand tapestry of existence. By transcending our ego's boundaries, we may experience the oneness of all being, becoming instruments of love, compassion, and wisdom.

As we embark on the river of renewal, we become part of its eternal flow, carrying wisdom's transformative power into our lives. The river's currents remind us that renewal is a lifelong journey, not a destination – a dynamic process of growth, transformation, and rebirth.

As we navigate the river's twists and turns, we encounter the waterfalls of awakening, where sudden insights illuminate our path. Wisdom's enlightened sages – from Buddha to Ramana Maharshi – demonstrate the power of awakening, transcending the veil of ignorance. By embracing awakening's transformative power, we may liberate ourselves from conditioning, recognizing our true nature.

The river's tranquil pools reflect the beauty of mindfulness, where presence and awareness guide our journey. Wisdom's mindful masters – from Thich Nhat Hanh to Jon Kabat-Zinn – show us how to cultivate mindfulness, embracing life's precious moments. By anchoring ourselves in the present, we may transcend the distractions of our minds, discovering peace amidst turmoil.

As we flow downstream, we encounter the river's confluences, where diverse perspectives merge into a unified whole. Wisdom's integrative thinkers – from Aristotle to Ken Wilber – demonstrate the power of synthesis, revealing the interconnectedness of all things. By embracing integration's wisdom, we may transcend fragmentation, addressing the complex challenges of our time.

The river's banks are lined with the gardens of gratitude, where appreciation and thankfulness nourish our souls. Wisdom's compassionate sages – from Meister Eckhart to Desmond Tutu

– show us how to cultivate gratitude, recognizing life's precious gifts. By embracing gratitude's transformative power, we may radiate love, compassion, and wisdom, illuminating the world.

As we approach the river's source, we encounter the misty veil of mystery, where the unknown beckons us forward. Wisdom's mystical explorers – from Rumi to Hafiz – demonstrate the power of embracing mystery, surrendering to the unknown. By transcending our need for certainty, we may discover the hidden patterns and rhythms governing reality.

The river of renewal reminds us that transformation is a lifelong journey, not a destination – a dynamic process of growth, decay, and rebirth. Wisdom's timeless sages whisper secrets to those who listen, guiding us toward the ultimate goal: self-realization.

As we merge with the river's flow, we become instruments of wisdom, carrying its transformative power into our lives. The river's currents remind us that renewal is not just a personal journey but a collective one, illuminating the path for all humanity.

9

Chapter 8:The ocean of oneness

As we embark on the ocean of oneness, we enter the vast expanse of unity, where wisdom's deepest currents connect us all. This boundless sea – navigated by sages, mystics, and visionaries – represents the ultimate reality, transcending borders, boundaries, and distinctions. Oneness's waves crash upon the shores of our hearts, reminding us of our shared humanity.

The ocean's surface reflects the beauty of diversity, where unique perspectives and experiences converge into a vibrant tapestry. Wisdom's inclusive sages – from Rumi to Desmond Tutu – demonstrate the power of embracing diversity, celebrating our differences. By honoring the richness of human experience, we may transcend fragmentation, recognizing our essential unity.

As we dive into the ocean's depths, we encounter the currents of compassion, where empathy and kindness guide our actions. Wisdom's compassionate leaders – from Buddha to Nelson Mandela – show us how to cultivate compassion, recognizing the interconnectedness of all beings. By embracing compassion's transformative power, we may heal the wounds of separation, radiating love and understanding.

The ocean's depths also conceal the treasures of intuition, where instinct and insight illuminate our path. Wisdom's intuitive guides – from Carl Jung to Clarissa Pinkola Estés – demonstrate the power of trusting our inner wisdom, listening to the whispers of our souls. By embracing intuition's guidance, we may navigate life's complexities, discovering hidden patterns and rhythms.

As we explore the ocean's mysteries, we encounter the ridges of interdependence, where individuality merges with the collective. Wisdom's holistic thinkers – from Buckminster Fuller to Fritjof Capra – show us how to recognize the intricate web of relationships binding our world. By embracing interdependence's wisdom, we may transcend ego's boundaries, addressing the complex challenges of our time.

The ocean's horizon beckons us toward the infinite possibilities of the human spirit, where creativity and innovation reshape our world. Wisdom's visionary artists – from Leonardo da Vinci to Frida Kahlo – demonstrate the power of imagination, converting dreams into reality. By embracing creativity's transformative power, we may birth new possibilities, illuminating the path for humanity.

As we merge with the ocean's oneness, we become instruments of wisdom, radiating love, compassion, and understanding. The ocean's waves remind us that unity is not just a destination but a journey – a dynamic process of growth, transformation, and connection.

The ocean of oneness whispers secrets to those who listen, guiding us toward the ultimate goal: self-realization. Wisdom's timeless sages remind us that our individual journeys are intertwined, forming the grand tapestry of human experience. By embracing the ocean's wisdom, we may become the embodiment of oneness, illuminating the world.As we navigate the ocean's depths, we encounter the underwater forests of forgiveness, where releasing the past frees us from its burdens. Wisdom's compassionate teachers – from Thich Nhat Hanh to Desmond Tutu – demonstrate the power of forgiveness, healing the wounds of yesterday. By embracing forgiveness's liberating power, we may release our attachment to pain, rediscovering our essential unity.

The ocean's currents carry us toward the shores of gratitude, where appreciation and thankfulness nourish our souls. Wisdom's grateful sages – from Meister Eckhart to Maya Angelou – show us how to cultivate gratitude, recognizing life's precious gifts. By embracing gratitude's transformative power, we may radiate love, compassion, and wisdom, illuminating the world.

As we explore the ocean's mysteries, we discover the hidden caves of humility, where self-awareness and modesty guide our journey. Wisdom's humble leaders – from Socrates to Nelson Mandela – demonstrate the power of humility, recognizing the limits of our knowledge. By embracing humility's wisdom, we may transcend ego's boundaries, embracing the interconnectedness of all beings.

The ocean's waves crash upon the rocks of resilience, where adaptability and determination overcome adversity. Wisdom's courageous innovators – from Steve Jobs to Malala Yousafzai – show us how to navigate uncertainty, converting obstacles into opportunities. By embracing resilience's transformative power, we may transcend our limitations, achieving greatness.

As we merge with the ocean's oneness, we become part of its eternal flow, carrying wisdom's transformative power into our lives. The ocean's tides remind us that unity is a dynamic process, not a static state – a continuous unfolding of growth, transformation, and connection.

The ocean of oneness holds the secrets of the universe, whispering truths to those who listen. Wisdom's timeless sages guide us toward the ultimate reality, transcending borders, boundaries, and distinctions. By embracing the ocean's wisdom, we may become the embodiment of unity, radiating love, compassion, and wisdom.

As we sail across the ocean's vast expanse, we encounter the islands of insight, where sudden understanding illuminates our path. Wisdom's enlightened sages – from Buddha to Einstein – demonstrate the power of insight, revealing the hidden patterns and rhythms governing reality. By embracing insight's transformative power, we may transcend ignorance, recognizing our essential unity.

10

Chapter 9:The temple of timelessness

As we approach the temple of timelessness, we enter the realm of the eternal, where wisdom's timeless principles guide us. This sacred sanctuary – revered by sages, philosophers, and visionaries – represents the ultimate reality, transcending the bounds of time and space. Timelessness's gates open to reveal the secrets of the ages.

The temple's entrance is flanked by the pillars of perspective, where broad vistas reveal the interconnectedness of all things. Wisdom's holistic thinkers – from Aristotle to Ken Wilber – demonstrate the power of synthesizing knowledge, recognizing the unity that underlies human experience. By embracing perspective's wisdom, we may transcend fragmentation, addressing the complex challenges of our time.

As we step into the temple's grand hall, we encounter the statues of significance, where purpose and meaning illuminate our lives. Wisdom's purpose-driven leaders – from Nelson Mandela to Jane Goodall – show us how to cultivate significance, aligning our actions with our values. By embracing significance's transformative power, we may leave a lasting legacy, inspiring future generations.

The temple's inner sanctum holds the sacred texts of self-awareness, where introspection and reflection guide our journey. Wisdom's contemplative sages – from Marcus Aurelius to Thich Nhat Hanh – demonstrate the power of self-awareness, confronting fears, desires, and limitations. By embracing self-awareness's liberating power, we may transcend ego's boundaries, recognizing our essential unity.

As we ascend to the temple's upper chambers, we encounter the windows of wonder, where awe and curiosity illuminate our understanding. Wisdom's visionary explorers – from Leonardo da Vinci to Carl Sagan – show us how to cultivate wonder, embracing the mysteries of existence. By embracing wonder's transformative power, we may transcend our limitations, discovering new worlds and possibilities.

The temple's pinnacle offers a panoramic view of the timeless landscape, where wisdom's eternal principles guide us. Wisdom's timeless sages remind us that timelessness is not just a destination but a journey – a dynamic process of growth, transformation, and connection.

The temple's foundations are rooted in the bedrock of compassion, where empathy and kindness guide our actions. Wisdom's compassionate leaders – from Buddha to Desmond Tutu – demonstrate the power of cultivating compassion, recognizing the interconnectedness of all beings.

CHAPTER 9:THE TEMPLE OF TIMELESSNESS

As we explore the temple's labyrinthine corridors, we discover the mirrors of mindfulness, where presence and awareness reflect our true nature. Wisdom's mindful masters – from Thich Nhat Hanh to Jon Kabat-Zinn – show us how to cultivate mindfulness, embracing life's precious moments.

The temple's ceiling is adorned with the frescoes of forgiveness, where releasing the past frees us from its burdens. Wisdom's forgiving sages – from Jesus to Gandhi – demonstrate the power of forgiveness, healing the wounds of yesterday.

As we delve deeper into the temple, we encounter the chambers of clarity, where precision and discernment guide our understanding. Wisdom's analytical thinkers – from René Descartes to Martha Nussbaum – demonstrate the power of critical thinking, evaluating evidence and challenging assumptions.

The temple's walls are adorned with the tapestries of tradition, where cultural heritage and collective memory enrich our lives. Wisdom's cultural guardians – from Confucius to Chimamanda Ngozi Adichie – show us how to honor our roots, embracing the wisdom of our ancestors.

As we ascend to the temple's upper levels, we encounter the observatories of openness, where curiosity and receptivity expand our horizons. Wisdom's visionary explorers – from Copernicus to Rachel Carson – demonstrate the power of embracing new ideas, challenging conventional wisdom.

The temple's sanctum sanctorum holds the sacred flame of inner wisdom, where intuition and instinct guide our journey.

Wisdom's intuitive guides – from Carl Jung to Clarissa Pinkola Estés – show us how to trust our inner voice, listening to the whispers of our souls.

The temple's surroundings are nourished by the gardens of gratitude, where appreciation and thankfulness cultivate our inner strength. Wisdom's grateful sages – from Epictetus to Mary Oliver – demonstrate the power of gratitude, recognizing life's precious gifts.

As we merge with the temple's timeless essence, we become instruments of wisdom, radiating clarity, compassion, and understanding. The temple's wisdom reminds us that timelessness is not just a destination but a journey – a dynamic process of growth, transformation, and connection.

The temple's architecture reflects the harmony of balance, where opposing forces are reconciled. Wisdom's balancing sages – from Lao Tzu to Hannah Arendt – demonstrate the power of finding equilibrium, navigating life's complexities.

Within the temple's heart lies the sanctuary of simplicity, where clarity and elegance reveal the essence of reality. Wisdom's simplicity advocates – from Henry David Thoreau to Dieter Rams – show us how to strip away unnecessary complexity, uncovering the beauty of simplicity.

The temple's treasury holds the jewels of justice, where fairness and compassion guide our actions. Wisdom's justice advocates – from Socrates to Ruth Bader Ginsburg – demonstrate the power of upholding justice, protecting the vulnerable.

As we explore the temple's labyrinthine passages, we discover the mirrors of mortality, where awareness of our finite existence focuses our priorities. Wisdom's mortality awareness teachers – from Montaigne to Atul Gawande – show us how to confront our own mortality, living with intention.

The temple's celestial maps chart the movements of the cosmos, reminding us of our place within the universe. Wisdom's cosmological thinkers – from Pythagoras to Neil deGrasse Tyson – demonstrate the power of contemplating the vastness of existence.

The temple's library contains the scrolls of storytelling, where narrative and mythology convey timeless truths. Wisdom's storytelling masters – from Homer to Toni Morrison – show us how to weave tales that inspire, educate, and transform.

As we depart the temple, we carry its timeless wisdom into our lives, applying its principles to navigate life's challenges. The temple's wisdom reminds us that timelessness is not just a concept but a living reality – a dynamic process of growth, transformation, and connection.

11

Chapter 10: The garden of enlightenment

As we enter the garden of enlightenment, we step into the realm of ultimate understanding, where wisdom's radiant flowers bloom. This serene oasis – cultivated by sages, mystics, and visionaries – represents the culmination of our journey, where the mysteries of existence are revealed.

The garden's entrance is flanked by the gates of awareness, where mindfulness and presence guide our passage. Wisdom's mindful masters – from Thich Nhat Hanh to Eckhart Tolle – demonstrate the power of living in the present, transcending the veil of ignorance.

As we stroll through the garden's lush pathways, we encounter the trees of transcendence, where ego's boundaries dissolve into unity. Wisdom's transcendent sages – from Rumi to Ramana Maharshi – show us how to surrender our attachments, merging with the divine.

CHAPTER 10: THE GARDEN OF ENLIGHTENMENT

The garden's central fountain represents the source of wisdom, where intuition and insight flow. Wisdom's intuitive guides – from Carl Jung to Clarissa Pinkola Estés – demonstrate the power of trusting our inner wisdom, listening to the whispers of our souls.

The garden's various flowers symbolize the facets of enlightenment: compassion, clarity, and inner peace. Wisdom's enlightened sages – from Buddha to Dalai Lama – demonstrate the power of cultivating these qualities, radiating love and light.

As we explore the garden's hidden corners, we discover the benches of reflection, where self-awareness and introspection guide our journey. Wisdom's reflective thinkers – from Socrates to Rousseau – show us how to examine our lives, confronting our fears and limitations.

The garden's pinnacle offers a panoramic view of the landscape of understanding, where wisdom's timeless principles guide us. Wisdom's enlightened sages remind us that enlightenment is not just a destination but a journey – a dynamic process of growth, transformation, and connection.

The garden's winding streams represent the flow of wisdom, nourishing our minds and hearts. Wisdom's fluid thinkers – from Heraclitus to Henri Bergson – demonstrate the power of embracing change, adapting to life's fluidity.

The garden's statues of silence remind us of the power of stillness, quieting the mind and listening to the heart. Wisdom's silent sages – from Pythagoras to Thomas Merton – show us how to cultivate inner stillness, accessing the depths of our being.

As we depart the garden, we carry its enlightened wisdom into our lives, applying its principles to navigate life's challenges. The garden's wisdom reminds us that enlightenment is not just a concept but a living reality – a dynamic process of growth, transformation, and connection.The garden's aromatic herbs symbolize the fragrance of gratitude, filling our lives with appreciation and thankfulness. Wisdom's grateful sages – from Epictetus to Mary Oliver – demonstrate the power of cultivating gratitude, recognizing life's precious gifts.

As we explore the garden's maze, we encounter the mirrors of self-reflection, where introspection and awareness guide our journey. Wisdom's reflective thinkers – from Montaigne to James Baldwin – show us how to examine our lives, confronting our fears and limitations.

The garden's sundial reminds us of the preciousness of time, urging us to live in the present. Wisdom's timeless sages – from Seneca to Alan Watts – demonstrate the power of transcending time's constraints, living in eternity.

The garden's celestial alignments represent the harmony of the universe, guiding us toward cosmic balance. Wisdom's cosmological thinkers – from Pythagoras to Brian Swimme – show us how to recognize the interconnectedness of all existence.

As we sit beneath the garden's ancient trees, we experience the

CHAPTER 10: THE GARDEN OF ENLIGHTENMENT

wisdom of aging, where life's experiences distill into insight. Wisdom's elder sages – from Confucius to Nelson Mandela – demonstrate the power of aging wisely, sharing their wisdom with humility.

The garden's hidden grottos contain the treasures of forgiveness, where releasing the past frees us from its burdens. Wisdom's forgiving sages – from Jesus to Desmond Tutu – show us how to cultivate forgiveness, healing the wounds of yesterday.

As we merge with the garden's essence, we become instruments of wisdom, radiating love, compassion, and understanding. The garden's wisdom reminds us that enlightenment is not just a destination but a journey – a dynamic process of growth, transformation, and connection.The garden's labyrinthine paths represent the journey of self-discovery, where every step reveals new insights. Wisdom's self-discovery guides – from Socrates to Joseph Campbell – demonstrate the power of exploring our inner depths, uncovering our true potential.

As we stroll through the garden's vibrant meadows, we encounter the flowers of creativity, where imagination and innovation bloom. Wisdom's creative visionaries – from Leonardo da Vinci to Maya Angelou – show us how to cultivate creativity, bringing new ideas to life.

The garden's tranquil ponds reflect the calmness of inner peace, where wisdom's gentle waters soothe our souls. Wisdom's peaceful sages – from Buddha to Dalai Lama – demonstrate the power of cultivating inner peace, radiating serenity.

The garden's majestic trees symbolize the strength of resilience, where adaptability and determination overcome adversity. Wisdom's resilient leaders – from Nelson Mandela to Malala Yousafzai – show us how to navigate uncertainty, emerging stronger.

As we explore the garden's hidden streams, we discover the sources of inspiration, where passion and purpose ignite our lives. Wisdom's inspiring mentors – from Aristotle to Oprah Winfrey – demonstrate the power of cultivating passion, living a life of purpose.

The garden's radiant sunlight represents the illumination of understanding, where wisdom's light dispels ignorance. Wisdom's enlightened sages – from Plato to Eckhart Tolle – show us how to cultivate understanding, recognizing the unity of all existence.

The garden's gentle breezes carry the whispers of wisdom, guiding us toward our highest potential. Wisdom's gentle guides – from Rumi to Mary Oliver – demonstrate the power of listening to our inner voice, trusting our intuition.

The garden's majestic architecture represents the grandeur of human potential, where wisdom's timeless principles guide us toward excellence. As we explore the garden's intricate designs, we discover the interconnectedness of all aspects of human existence, from the smallest detail to the vast expanse of our collective experience. Wisdom's holistic thinkers – from Aristotle to Ken Wilber – demonstrate the power of synthesizing knowledge, recognizing the unity that underlies human understanding.

As we wander through the garden's serene landscapes, we encounter the lakes of introspection, where self-awareness and reflection guide our journey. Wisdom's introspective sages – from Marcus Aurelius to James Baldwin – show us how to cultivate self-awareness, confronting our fears and limitations, and emerging stronger and wiser. The garden's lakes also symbolize the depth of human emotion, where compassion and empathy flow like gentle streams, nourishing our hearts and souls.

The garden's winding paths represent the journey of personal growth, where every step reveals new insights and challenges. Wisdom's growth-oriented leaders – from Socrates to Nelson Mandela – demonstrate the power of embracing challenge, transcending limitations, and emerging stronger and more resilient. As we navigate the garden's twists and turns, we discover the value of perseverance, adaptability, and determination, essential qualities for navigating life's complexities.

The garden's vibrant flora represents the diversity of human experience, where unique perspectives and talents bloom like flowers in a kaleidoscope of colors. Wisdom's celebrants of diversity – from Confucius to Chimamanda Ngozi Adichie – show us how to honor our differences, embracing the richness of human culture and experience. The garden's flora also symbolizes the power of creativity, where imagination and innovation flourish, bringing new ideas and possibilities to life.

As we explore the garden's hidden treasures, we discover the jewels of wisdom, where timeless principles guide us toward enlightened living. Wisdom's enlightened sages – from Buddha to Eckhart Tolle – demonstrate the power of cultivating mindfulness, compassion, and understanding, radiating love and light. The garden's treasures also represent the wealth of human knowledge, where wisdom's timeless principles await discovery, guiding us toward our highest potential.

The garden's majestic waterfalls represent the transformative power of wisdom, cascading into the depths of our being. Wisdom's transformative guides – from Rumi to Paulo Coelho – demonstrate the power of embracing change, surrendering to the flow of life. As we stand beneath the falls, we experience the cleansing of our minds and hearts, washing away fears, doubts, and limitations.

The garden's luminous stars symbolize the celestial guidance of wisdom, illuminating our path through life's complexities. Wisdom's celestial navigators – from Pythagoras to Carl Sagan – show us how to recognize the harmony of the universe, aligning our lives with cosmic principles. The stars also represent the infinite possibilities of human potential, shining brightly like beacons in the night sky.

As we explore the garden's hidden grottos, we discover the ancient wisdom of the ages, inscribed on walls of timeless stone. Wisdom's ancient sages – from Hermes Trismegistus to Lao Tzu – demonstrate the power of timeless principles, guiding us toward enlightenment. The grottos also represent the sanctuaries of solitude, where introspection and contemplation nurture our souls.

CHAPTER 10: THE GARDEN OF ENLIGHTENMENT

The garden's resplendent rainbow bridges represent the connections between diverse perspectives, uniting humanity in a kaleidoscope of colors. Wisdom's bridge-builders – from Martin Luther King Jr. to Malala Yousafzai – show us how to cultivate empathy, compassion, and understanding, transcending boundaries of culture, creed, and nationality.

The garden's majestic mountains symbolize the towering achievements of human wisdom, where pinnacle moments of insight and understanding reshape our lives. Wisdom's mountaineers – from Aristotle to Albert Einstein – demonstrate the power of scaling new heights, pushing the boundaries of human knowledge and understanding.

As we merge with the garden's essence, we become instruments of wisdom, radiating love, compassion, and understanding. The garden's wisdom reminds us that enlightenment is not just a destination but a journey – a dynamic process of growth, transformation, and connection.

12

Chapter 11:The Oasis on inner peace

As we journey through the arid landscapes of life, we arrive at the oasis of inner peace, where wisdom's refreshing waters quench our thirst for serenity. This serene sanctuary – cultivated by sages, mystics, and visionaries – represents the ultimate refuge from life's turmoil.

The oasis's entrance is flanked by the palms of patience, where calmness and understanding guide our passage. Wisdom's patient sages – from Epictetus to Dalai Lama – demonstrate the power of embracing patience, navigating life's challenges with equanimity.

As we stroll through the oasis's lush gardens, we encounter the fountains of forgiveness, where releasing the past frees us from its burdens. Wisdom's forgiving sages – from Jesus to Desmond Tutu – show us how to cultivate forgiveness, healing the wounds of yesterday.

The oasis's central pavilion represents the hub of inner peace,

CHAPTER 11: THE OASIS ON INNER PEACE

where mindfulness and meditation converge. Wisdom's mindful masters – from Buddha to Thich Nhat Hanh – demonstrate the power of cultivating mindfulness, anchoring our lives in the present.

The oasis's tranquil lakes reflect the calmness of our inner world, where wisdom's gentle waters soothe our souls. Wisdom's peaceful sages – from Rumi to Mary Oliver – show us how to cultivate inner peace, radiating serenity.

As we explore the oasis's flowering gardens, we discover the blooms of gratitude, where appreciation and thankfulness nourish our hearts. Wisdom's grateful sages – from Marcus Aurelius to Oprah Winfrey – demonstrate the power of cultivating gratitude, recognizing life's precious gifts.

The oasis's majestic date trees symbolize the wisdom of aging, where life's experiences distill into insight. Wisdom's elder sages – from Confucius to Nelson Mandela – demonstrate the power of aging wisely, sharing their wisdom with humility. The oasis's serene architecture represents the harmony of body, mind, and spirit, where wisdom's timeless principles guide our well-being. Wisdom's holistic thinkers – from Aristotle to Deepak Chopra – demonstrate the power of integrating physical, emotional, and spiritual health.

As we wander through the oasis's fragrant orchards, we encounter the trees of trust, where faith and confidence flourish. Wisdom's trusting sages – from Socrates to Martin Luther King Jr. – show us how to cultivate trust, relying on our inner guidance.

The oasis's shimmering mirages symbolize the illusions of the ego, where wisdom's clarity dispels deception. Wisdom's illuminating sages – from Buddha to Eckhart Tolle – demonstrate the power of transcending ego's limitations, revealing our true nature.

The oasis's ancient artifacts represent the wisdom of the ages, where timeless principles await discovery. Wisdom's preservationists – from Pythagoras to Joseph Campbell – show us how to honor our heritage, learning from the past.

As we rest beneath the oasis's shaded palms, we experience the solace of silence, where wisdom's quiet whispers guide our hearts. Wisdom's silent sages – from Quakers to Sufi mystics – demonstrate the power of embracing silence, listening to our inner voice.

The oasis's celestial alignments represent the harmony of heaven and earth, where wisdom's cosmic principles guide our lives. Wisdom's cosmological thinkers – from Copernicus to Brian Swimme – show us how to recognize our place within the universe.

The oasis's gentle breezes carry the whispers of wisdom, guiding us toward our highest potential. Wisdom's gentle guides – from Rumi to Mary Oliver – demonstrate the power of listening to our inner voice.

The oasis's resplendent sunsets symbolize the beauty of impermanence, where wisdom's timeless principles transcend change. Wisdom's impermanence teachers – from Heraclitus to Alan

Watts – show us how to embrace the fluidity of life.

As we depart the oasis, we carry its wisdom into our lives, applying its principles to navigate life's challenges. The oasis's wisdom reminds us that inner peace is not just a destination but a journey – a dynamic process of growth, transformation, and connection.The oasis's luminous stars represent the celestial navigation of wisdom, guiding us through life's uncertainties. Wisdom's stellar navigators – from Pythagoras to Carl Sagan – demonstrate the power of recognizing our place within the universe.

As we explore the oasis's hidden springs, we discover the sources of renewal, where wisdom's rejuvenating waters revitalize our spirits. Wisdom's renewing sages – from Lao Tzu to Rachel Carson – show us how to cultivate renewal, embracing the cycles of life.

The oasis's majestic mountains symbolize the towering achievements of human resilience, where wisdom's timeless principles guide us through adversity. Wisdom's resilient leaders – from Nelson Mandela to Malala Yousafzai – demonstrate the power of overcoming obstacles.

The oasis's vibrant marketplaces represent the exchange of wisdom, where diverse perspectives enrich our understanding. Wisdom's marketplace sages – from Socrates to Paulo Coelho – show us how to cultivate dialogue, embracing the richness of human experience.

As we wander through the oasis's fragrant gardens, we encounter the flowers of forgiveness, where releasing the past frees us from its burdens. Wisdom's forgiving sages – from Jesus to Desmond Tutu – demonstrate the power of cultivating forgiveness.

The oasis's ancient manuscripts represent the wisdom of the ages, where timeless principles await discovery. Wisdom's preservationists – from Confucius to Joseph Campbell – show us how to honor our heritage.

The oasis's shimmering mirages symbolize the illusions of perception, where wisdom's clarity dispels deception. Wisdom's illuminating sages – from Buddha to Eckhart Tolle – demonstrate the power of transcending limitation.The oasis's tranquil atmosphere represents the serenity of wisdom, calming the mind and soothing the soul. Wisdom's serene sages – from Epictetus to Dalai Lama – demonstrate the power of cultivating inner peace.

As we explore the oasis's majestic architecture, we discover the halls of self-reflection, where introspection guides our journey. Wisdom's reflective thinkers – from Socrates to James Baldwin – show us how to examine our lives.

The oasis's resplendent fountains symbolize the flow of creativity, where imagination and innovation nourish our spirits. Wisdom's creative visionaries – from Leonardo da Vinci to Maya Angelou – demonstrate the power of cultivating creativity.

The oasis's lush gardens represent the growth of wisdom, where patience and nurturing guide our development. Wisdom's gardening sages – from Lao Tzu to Rachel Carson – show us how to cultivate wisdom.

As we stroll through the oasis's vibrant streets, we encounter the marketplaces of ideas, where diverse perspectives enrich our understanding. Wisdom's marketplace sages – from Aristotle to Paulo Coelho – demonstrate the power of dialogue.

The oasis's ancient artifacts represent the wisdom of the ages, where timeless principles await discovery. Wisdom's preservationists – from Pythagoras to Joseph Campbell – show us how to honor our heritage.

The oasis's celestial music represents the harmony of the universe, where wisdom's cosmic principles guide our lives. Wisdom's cosmological thinkers – from Copernicus to Brian Swimme – demonstrate the power of recognizing our place.The oasis's radiant sunlight symbolizes the illumination of understanding, where wisdom's light dispels ignorance. Wisdom's enlightening sages – from Buddha to Eckhart Tolle – demonstrate the power of cultivating insight.

The oasis's gentle showers represent the nourishment of wisdom, refreshing our minds and hearts. Wisdom's nourishing sages – from Confucius to Mary Oliver – show us how to cultivate gratitude.

As we depart the oasis, we carry its wisdom into our lives, applying its principles. The oasis's wisdom reminds us inner peace is a journey – growth, transformation, connection.

13

Chapter 12: The summit of enlightenment

As we reach the summit of enlightenment, we behold the breathtaking panorama of wisdom, where timeless principles guide us toward ultimate understanding. This majestic pinnacle – cultivated by sages, mystics, and visionaries – represents the culmination of our journey. The summit's entrance is flanked by the gates of self-realization, where introspection and awareness guide our passage. Wisdom's self-realization sages – from Socrates to Ramana Maharshi – demonstrate the power of recognizing our true nature.

As we ascend the summit's winding paths, we encounter the mirrors of mindfulness, where presence and awareness reflect our essence. Wisdom's mindful masters – from Buddha to Thich Nhat Hanh – show us how to cultivate mindfulness.

The summit's central peak represents the pinnacle of enlightenment, where wisdom's radiant light illuminates our lives. Wisdom's enlightened sages – from Jesus to Eckhart Tolle –

CHAPTER 12: THE SUMMIT OF ENLIGHTENMENT

demonstrate the power of embodying wisdom.The summit's resplendent sunrise symbolizes the dawn of awareness, where wisdom's light dispels ignorance. Wisdom's illuminating sages – from Plato to Alan Watts – demonstrate the power of cultivating insight.

As we explore the summit's majestic landscapes, we discover the valleys of compassion, where empathy and understanding nourish our hearts. Wisdom's compassionate sages – from Dalai Lama to Maya Angelou – show us how to cultivate compassion.

The summit's ancient artifacts represent the wisdom of the ages, where timeless principles await discovery. Wisdom's preservationists – from Confucius to Joseph Campbell – demonstrate the power of honoring our heritage.

The summit's celestial music represents the harmony of the universe, where wisdom's cosmic principles guide our lives. Wisdom's cosmological thinkers – from Pythagoras to Brian Swimme – demonstrate the power of recognizing our place.

As we behold the summit's breathtaking view, we realize the interconnectedness of all existence. Wisdom's unity sages – from Rumi to Tagore – demonstrate the power of embracing unity.The summit's radiant aura represents the illumination of wisdom, guiding us toward our highest potential. Wisdom's illuminating sages – from Socrates to Eckhart Tolle – demonstrate the power of cultivating insight.

As we explore the summit's hidden caverns, we discover the treasures of intuition, where inner wisdom guides our decisions.

Wisdom's intuitive guides – from Carl Jung to Clarissa Pinkola Estés – show us how to trust our inner voice.

The summit's majestic waterfalls symbolize the transformative power of wisdom, cascading into the depths of our being. Wisdom's transformative sages – from Rumi to Paulo Coelho – demonstrate the power of embracing change.

The summit's lush gardens represent the growth of wisdom, where patience and nurturing guide our development. Wisdom's gardening sages – from Lao Tzu to Rachel Carson – show us how to cultivate wisdom.

As we stroll through the summit's vibrant marketplaces, we encounter the exchange of ideas, where diverse perspectives enrich our understanding. Wisdom's marketplace sages – from Aristotle to Paulo Coelho – demonstrate the power of dialogue.

The summit's ancient manuscripts represent the wisdom of the ages, where timeless principles await discovery. Wisdom's preservationists – from Pythagoras to Joseph Campbell – demonstrate the power of honoring our heritage.

The summit's resplendent rainbow bridges represent the connections between diverse perspectives, uniting humanity in a kaleidoscope of colors. Wisdom's bridge-builders – from Martin Luther King Jr. to Malala Yousafzai – show us how to cultivate empathy.The summit's celestial alignments represent the harmony of heaven and earth, where wisdom's cosmic principles guide our lives. Wisdom's cosmological thinkers – from Copernicus to Brian Swimme – demonstrate the power of recognizing our place.

As we behold the summit's breathtaking view, we realize the interconnectedness of all existence. Wisdom's unity sages – from Rumi to Tagore – demonstrate the power of embracing unity.

The summit's gentle breezes carry the whispers of wisdom, guiding us toward our highest potential. Wisdom's gentle guides – from Mary Oliver to Desmond Tutu – demonstrate the power of listening.

The summit's radiant sunlight symbolizes the illumination of understanding, where wisdom's light dispels ignorance. Wisdom's enlightening sages – from Buddha to Eckhart Tolle – demonstrate the power of cultivating insight.The summit's majestic architecture represents the grandeur of human potential, where wisdom's timeless principles guide us toward excellence. Wisdom's visionary architects – from Leonardo da Vinci to Frank Lloyd Wright – demonstrate the power of designing a life of purpose.

As we explore the summit's hidden libraries, we discover the ancient tomes of wisdom, where timeless knowledge awaits discovery. Wisdom's preservationists – from Alexandria's librarians to Joseph Campbell – demonstrate the power of honoring our intellectual heritage.

The summit's resplendent art gallery represents the beauty of creative expression, where imagination and innovation nourish our souls. Wisdom's creative visionaries – from Michelangelo to

Maya Angelou – demonstrate the power of cultivating creativity.

The summit's vibrant community represents the unity of humanity, where diverse perspectives enrich our understanding. Wisdom's community builders – from Martin Luther King Jr. to Malala Yousafzai – demonstrate the power of cultivating empathy.

As we ascend the summit's winding paths, we encounter the mirrors of self-reflection, where introspection guides our journey. Wisdom's reflective thinkers – from Socrates to James Baldwin – demonstrate the power of examining our lives.

The summit's celestial music represents the harmony of the universe, where wisdom's cosmic principles guide our lives. Wisdom's cosmological thinkers – from Pythagoras to Brian Swimme – demonstrate the power of recognizing our place.

The summit's radiant aura represents the illumination of wisdom, guiding us toward our highest potential. Wisdom's illuminating sages – from Buddha to Eckhart Tolle – demonstrate the power of cultivating insight.The summit's gentle showers represent the nourishment of wisdom, refreshing our minds and hearts. Wisdom's nourishing sages – from Confucius to Mary Oliver – demonstrate the power of cultivating gratitude.

The summit's majestic mountains symbolize the towering achievements of human resilience, where wisdom's timeless principles guide us through adversity. Wisdom's resilient leaders – from Nelson Mandela to Malala Yousafzai – demonstrate the power of overcoming obstacles.

CHAPTER 12: THE SUMMIT OF ENLIGHTENMENT

As we behold the summit's breathtaking view, we realize the interconnectedness of all existence. Wisdom's unity sages – from Rumi to Tagore – demonstrate the power of embracing unity.The summit's luminous stars represent the celestial guidance of wisdom, illuminating our path through life's complexities. Wisdom's stellar navigators – from Pythagoras to Carl Sagan – demonstrate the power of recognizing our place.

As we explore the summit's vibrant tapestry, we discover the intricate threads of interconnection, weaving humanity into a single fabric. Wisdom's tapestry weavers – from Rumi to Maya Angelou – demonstrate the power of cultivating empathy.

The summit's resplendent rainbow bridges represent the connections between diverse perspectives, uniting humanity in a kaleidoscope of colors. Wisdom's bridge-builders – from Martin Luther King Jr. to Malala Yousafzai – show us how to cultivate understanding.

The summit's ancient artifacts represent the wisdom of the ages, where timeless principles await discovery. Wisdom's preservationists – from Confucius to Joseph Campbell – demonstrate the power of honoring our heritage.

As we ascend the summit's winding paths, we encounter the mirrors of self-awareness, where introspection guides our journey. Wisdom's reflective thinkers – from Socrates to James Baldwin – demonstrate the power of examining our lives.

The summit's majestic architecture represents the grandeur of human potential, where wisdom's timeless principles guide us toward excellence. Wisdom's visionary architects – from Leonardo da Vinci to Frank Lloyd Wright – demonstrate the power of designing a life of purpose.

The summit's radiant sunlight symbolizes the illumination of understanding, where wisdom's light dispels ignorance. Wisdom's enlightening sages – from Buddha to Eckhart Tolle – demonstrate the power of cultivating insight.The summit's gentle breezes carry the whispers of wisdom, guiding us toward our highest potential. Wisdom's gentle guides – from Mary Oliver to Desmond Tutu – demonstrate the power of listening.

The summit's luminous horizon represents the limitless possibilities of human potential, where wisdom's timeless principles guide us toward excellence. Wisdom's visionary leaders – from Nelson Mandela to Malala Yousafzai – demonstrate the power of cultivating hope.

As we behold the summit's breathtaking view, we realize the interconnectedness of all existence. Wisdom's unity sages – from Rumi to Tagore – demonstrate the power of embracing unity.

14

Chapter 13:The sanctuary of legacy

As we enter the sanctuary of legacy, we honor the timeless wisdom of the past, where the legacies of sages, visionaries, and leaders guide us toward a brighter future.The sanctuary's entrance is flanked by the gates of remembrance, where gratitude and appreciation guide our passage. Wisdom's remembrance sages – from Confucius to Maya Angelou – demonstrate the power of honoring our heritage.

As we explore the sanctuary's hallowed halls, we discover the chambers of reflection, where introspection and self-awareness illuminate our journey. Wisdom's reflective thinkers – from Socrates to James Baldwin – demonstrate the power of examining our lives.

The sanctuary's central shrine represents the pinnacle of legacy, where wisdom's timeless principles inspire future generations. Wisdom's legacy builders – from Leonardo da Vinci to Nelson Mandela – demonstrate the power of cultivating a lasting impact.The sanctuary's resplendent stained glass windows represent the kaleidoscope of human experience, where diverse perspectives enrich our understanding. Wisdom's stained glass artisans – from Michelangelo to Matisse – demonstrate the power of cultivating creativity.

As we wander through the sanctuary's serene gardens, we encounter the fountains of gratitude, where appreciation and thankfulness nourish our souls. Wisdom's grateful sages – from Marcus Aurelius to Oprah Winfrey – demonstrate the power of cultivating gratitude.

The sanctuary's ancient artifacts represent the wisdom of the ages, where timeless principles await discovery. Wisdom's preservationists – from Pythagoras to Joseph Campbell – demonstrate the power of honoring our intellectual heritage.

The sanctuary's majestic architecture represents the grandeur of human potential, where wisdom's timeless principles guide us toward excellence. Wisdom's visionary architects – from Frank Lloyd Wright to Zaha Hadid – demonstrate the power of designing a life of purpose.The sanctuary's luminous lanterns symbolize the guidance of wisdom, illuminating our path through life's complexities. Wisdom's lantern bearers – from Buddha to Eckhart Tolle – demonstrate the power of cultivating insight.As we behold the sanctuary's breathtaking view, we realize the interconnectedness of all existence. Wisdom's unity sages – from Rumi to Tagore – demonstrate the power of embracing unity.

The sanctuary's gentle breezes carry the whispers of wisdom, guiding us toward our highest potential. Wisdom's gentle guides – from Mary Oliver to Desmond Tutu – demonstrate the power of listening. The sanctuary's radiant mosaics represent the diversity of human experience, where individual stories form a larger tapestry of understanding. Wisdom's mosaic artisans – from Byzantine masters to contemporary artists – demonstrate the power of cultivating diversity.

As we explore the sanctuary's hidden archives, we discover the scrolls of forgotten wisdom, where ancient knowledge awaits rediscovery. Wisdom's archivists – from Alexandria's librarians to modern-day preservationists – demonstrate the power of honoring our cultural heritage.

The sanctuary's resplendent frescoes represent the evolution of human thought, where wisdom's timeless principles guide us toward progress. Wisdom's fresco artists – from Michelangelo to Diego Rivera – demonstrate the power of cultivating creativity.

The sanctuary's peaceful courtyards symbolize the serenity of wisdom, calming the mind and soothing the soul. Wisdom's peaceful sages – from Epictetus to Dalai Lama – demonstrate the power of cultivating inner peace.

As we ascend the sanctuary's winding staircases, we encounter the mirrors of self-awareness, where introspection guides our journey. Wisdom's reflective thinkers – from Socrates to James Baldwin – demonstrate the power of examining our lives.

The sanctuary's majestic domes represent the unity of humanity, where diverse perspectives converge beneath a shared canopy.

Wisdom's unity builders – from Martin Luther King Jr. to Malala Yousafzai – demonstrate the power of cultivating empathy.

The sanctuary's luminous stars symbolize the celestial guidance of wisdom, illuminating our path through life's complexities. Wisdom's stellar navigators – from Pythagoras to Carl Sagan – demonstrate the power of recognizing our place. The sanctuary's gentle showers represent the nourishment of wisdom, refreshing our minds and hearts. Wisdom's nourishing sages – from Confucius to Mary Oliver – demonstrate the power of cultivating gratitude.

The sanctuary's vibrant tapestries represent the intricate threads of interconnection, weaving humanity into a single fabric. Wisdom's tapestry weavers – from Rumi to Maya Angelou – demonstrate the power of cultivating empathy. As we behold the sanctuary's breathtaking view, we realize the interconnectedness of all existence. Wisdom's unity sages – from Rumi to Tagore – demonstrate the power of embracing unity.

The sanctuary's resplendent chandeliers symbolize the illumination of wisdom, casting light upon our path. Wisdom's illuminating sages – from Buddha to Eckhart Tolle – demonstrate the power of cultivating insight.

As we explore the sanctuary's hidden chambers, we discover the treasures of forgotten wisdom, where ancient knowledge awaits rediscovery. Wisdom's treasure hunters – from Herodotus to Joseph Campbell – demonstrate the power of honoring our cultural heritage.

CHAPTER 13: THE SANCTUARY OF LEGACY

The sanctuary's peaceful gardens represent the serenity of wisdom, calming the mind and soothing the soul. Wisdom's peaceful sages – from Epictetus to Dalai Lama – demonstrate the power of cultivating inner peace.

The sanctuary's majestic architecture represents the grandeur of human potential, where wisdom's timeless principles guide us toward excellence. Wisdom's visionary architects – from Frank Lloyd Wright to Zaha Hadid – demonstrate the power of designing a life of purpose.

As we ascend the sanctuary's winding staircases, we encounter the mirrors of self-awareness, where introspection guides our journey. Wisdom's reflective thinkers – from Socrates to James Baldwin – demonstrate the power of examining our lives.

The sanctuary's luminous stained glass windows represent the kaleidoscope of human experience, where diverse perspectives enrich our understanding. Wisdom's stained glass artisans – from Michelangelo to Matisse – demonstrate the power of cultivating creativity.

The sanctuary's ancient artifacts represent the wisdom of the ages, where timeless principles await discovery. Wisdom's preservationists – from Pythagoras to Joseph Campbell – demonstrate the power of honoring our intellectual heritage.The sanctuary's gentle breezes carry the whispers of wisdom, guiding us toward our highest potential. Wisdom's gentle guides – from Mary Oliver to Desmond Tutu – demonstrate the power of listening.

The sanctuary's radiant mosaics represent the diversity of human experience, where individual stories form a larger tapestry of understanding. Wisdom's mosaic artisans – from Byzantine masters to contemporary artists – demonstrate the power of cultivating diversity.

As we behold the sanctuary's breathtaking view, we realize the interconnectedness of all existence. Wisdom's unity sages – from Rumi to Tagore – demonstrate the power of embracing unity.The sanctuary's resplendent frescoes represent the evolution of human thought, where wisdom's timeless principles guide us toward progress. Wisdom's fresco artists – from Michelangelo to Diego Rivera – demonstrate the power of cultivating creativity.

The sanctuary's peaceful courtyards symbolize the serenity of wisdom, calming the mind and soothing the soul. Wisdom's peaceful sages – from Epictetus to Dalai Lama – demonstrate the power of cultivating inner peace.

The sanctuary's majestic domes represent the unity of humanity, where diverse perspectives converge beneath a shared canopy. Wisdom's unity builders – from Martin Luther King Jr. to Malala Yousafzai – demonstrate the power of cultivating empathy.

15

Chapter 14:The horizon of possibility

As we approach the horizon of possibility, we behold the vast expanse of human potential, where wisdom's timeless principles guide us toward excellence.The horizon's entrance is flanked by the gates of imagination, where creativity and innovation unlock our potential. Wisdom's imaginative sages – from Leonardo da Vinci to Maya Angelou – demonstrate the power of cultivating creativity.

As we explore the horizon's vast landscapes, we discover the mountains of resilience, where wisdom's timeless principles guide us through adversity. Wisdom's resilient leaders – from Nelson Mandela to Malala Yousafzai – demonstrate the power of overcoming obstacles.

The horizon's central peak represents the pinnacle of human achievement, where wisdom's timeless principles inspire future generations. Wisdom's visionary leaders – from Martin Luther King Jr. to Jane Goodall – demonstrate the power of cultivating a lasting impact.The horizon's radiant sunrise symbolizes the dawn of new possibilities, where wisdom's light dispels ignorance. Wisdom's illuminating sages – from Buddha to Eckhart Tolle – demonstrate the power of cultivating insight.

As we wander through the horizon's rolling hills, we encounter the gardens of gratitude, where appreciation and thankfulness nourish our souls. Wisdom's grateful sages – from Marcus Aurelius to Oprah Winfrey – demonstrate the power of cultivating gratitude.

The horizon's majestic architecture represents the grandeur of human potential, where wisdom's timeless principles guide us toward excellence. Wisdom's visionary architects – from Frank Lloyd Wright to Zaha Hadid – demonstrate the power of designing a life of purpose.

The horizon's luminous stars symbolize the celestial guidance of wisdom, illuminating our path through life's complexities. Wisdom's stellar navigators – from Pythagoras to Carl Sagan – demonstrate the power of recognizing our place.The horizon's gentle breezes carry the whispers of wisdom, guiding us toward our highest potential. Wisdom's gentle guides – from Mary Oliver to Desmond Tutu – demonstrate the power of listening.

The horizon's resplendent rainbows represent the diversity of human experience, where individual stories form a larger tapestry of understanding. Wisdom's rainbow weavers – from Rumi to Maya Angelou – demonstrate the power of cultivating empathy.

As we behold the horizon's breathtaking view, we realize the interconnectedness of all existence. Wisdom's unity sages – from Rumi to Tagore – demonstrate the power of embracing unity.The horizon's majestic mountains symbolize the towering achievements of human resilience, where wisdom's timeless principles guide us through adversity. Wisdom's resilient leaders – from Nelson Mandela to Malala Yousafzai – demonstrate the power of overcoming obstacles.

As we explore the horizon's hidden oases, we discover the treasures of forgotten wisdom, where ancient knowledge awaits rediscovery. Wisdom's treasure hunters – from Herodotus to Joseph Campbell – demonstrate the power of honoring our cultural heritage.

The horizon's radiant stars represent the celestial guidance of wisdom, illuminating our path through life's complexities. Wisdom's stellar navigators – from Pythagoras to Carl Sagan – demonstrate the power of recognizing our place.

The horizon's peaceful lakes symbolize the serenity of wisdom, calming the mind and soothing the soul. Wisdom's peaceful sages – from Epictetus to Dalai Lama – demonstrate the power of cultivating inner peace.

As we ascend the horizon's winding staircases, we encounter the mirrors of self-awareness, where introspection guides our journey. Wisdom's reflective thinkers – from Socrates to James Baldwin – demonstrate the power of examining our lives.

The horizon's luminous auroras represent the breathtaking beauty of wisdom, illuminating our understanding. Wisdom's aurora gazers – from Aristotle to Rachel Carson – demonstrate the power of cultivating wonder.

The horizon's resplendent archipelagos represent the diversity of human experience, where individual stories form a larger tapestry of understanding. Wisdom's archipelago explorers – from Homer to Maya Angelou – demonstrate the power of cultivating empathy.The horizon's gentle morning dew represents the refreshing power of wisdom, nourishing our minds and hearts. Wisdom's gentle guides – from Mary Oliver to Desmond Tutu – demonstrate the power of listening.

The horizon's vibrant marketplaces represent the exchange of ideas, where diverse perspectives enrich our understanding. Wisdom's marketplace sages – from Aristotle to Paulo Coelho – demonstrate the power of cultivating dialogue.As we behold the horizon's breathtaking view, we realize the interconnectedness of all existence. Wisdom's unity sages – from Rumi to Tagore – demonstrate the power of embracing unity.

The horizon's majestic mountains stood as testaments to the transformative power of wisdom, their rugged peaks and verdant valleys a reminder that even the most formidable challenges could be overcome through the application of timeless principles. Wisdom's resilient leaders, from Nelson Mandela to Malala Yousafzai, had demonstrated the power of overcoming obstacles, their unwavering commitment to justice and equality inspiring generations to strive for a brighter future.

As we explored the horizon's hidden oases, we discovered the treasures of forgotten wisdom, ancient knowledge that awaited rediscovery like a parched traveler stumbling upon a verdant oasis in the desert. Wisdom's treasure hunters, from Herodotus to Joseph Campbell, had spent their lives unearthing these hidden riches, their tireless efforts a testament to the enduring power of human curiosity.

The horizon's radiant stars shone like diamonds in the velvet expanse, their celestial guidance illuminating our path through life's complexities like a beacon shining brightly in the darkness. Wisdom's stellar navigators, from Pythagoras to Carl Sagan, had charted the cosmos, their groundbreaking discoveries expanding our understanding of the universe and our place within it.

The horizon's peaceful lakes sparkled like sapphires in the morning light, their serene waters calming the mind and soothing the soul like a gentle breeze rustling the leaves of a summer forest. Wisdom's peaceful sages, from Epictetus to Dalai Lama, had demonstrated the power of cultivating inner peace, their teachings a reminder that even in the midst of turmoil, tranquility could be found.

As we ascended the horizon's winding staircases, we encountered the mirrors of self-awareness, their reflective surfaces revealing the depths of our own understanding like a still pond reflecting the beauty of the surrounding landscape. Wisdom's reflective thinkers, from Socrates to James Baldwin, had gazed into these mirrors, their introspective journeys a testament to the power of examining our lives.

The horizon's luminous auroras danced across the sky like

ethereal curtains, their breathtaking beauty illuminating our understanding like a masterpiece of art unveiling the secrets of the human experience. Wisdom's aurora gazers, from Aristotle to Rachel Carson, had marveled at this beauty, their wonder-inspired discoveries expanding our understanding of the natural world.

The horizon's resplendent archipelagos stretched out before us like a tapestry of diverse perspectives, each island a unique thread in the intricate fabric of human experience. Wisdom's archipelago explorers, from Homer to Maya Angelou, had navigated these waters, their stories a testament to the power of cultivating empathy.

As we beheld the horizon's breathtaking view, we realized the interconnectedness of all existence, the boundaries between self and other dissolving like mist in the morning sun. Wisdom's unity sages, from Rumi to Tagore, had demonstrated the power of embracing unity, their teachings a reminder that we were not isolated individuals, but threads in the intricate tapestry of humanity.

16

Chapter 15:The haven of higher understanding

As we enter the haven of higher understanding, we discover the treasures of wisdom, where timeless principles guide us toward profound insight and inner peace.

The haven's entrance is flanked by the gates of introspection, where self-awareness and reflection illuminate our journey. Wisdom's reflective thinkers – from Socrates to James Baldwin – demonstrate the power of examining our lives.

As we explore the haven's vast library, we encounter the tomes of ancient wisdom, where the collective knowledge of humanity awaits discovery. Wisdom's sage scholars – from Confucius to Joseph Campbell – demonstrate the power of honoring our cultural heritage.

The haven's central chamber represents the pinnacle of higher understanding, where wisdom's timeless principles inspire profound insight and inner peace. Wisdom's enlightened sages

– from Buddha to Eckhart Tolle – demonstrate the power of cultivating mindfulness. The haven's radiant sunlight symbolizes the illumination of understanding, where wisdom's light dispels ignorance. Wisdom's enlightening sages – from Aristotle to Paulo Coelho – demonstrate the power of cultivating insight.

As we wander through the haven's serene walkways, we encounter the statues of compassion, where empathy and kindness guide our interactions. Wisdom's compassionate sages – from Dalai Lama to Maya Angelou – demonstrate the power of cultivating empathy.

The haven's peaceful lakes represent the serenity of wisdom, calming the mind and soothing the soul. Wisdom's peaceful sages – from Epictetus to Thich Nhat Hanh – demonstrate the power of cultivating inner peace.

The haven's majestic architecture represents the grandeur of human potential, where wisdom's timeless principles guide us toward excellence. Wisdom's visionary architects – from Frank Lloyd Wright to Zaha Hadid – demonstrate the power of designing a life of purpose. The haven's vibrant marketplace represents the exchange of ideas, where diverse perspectives enrich our understanding. Wisdom's marketplace sages – from Socrates to Paulo Coelho – demonstrate the power of cultivating dialogue.

As we behold the haven's breathtaking view, we realize the interconnectedness of all existence. Wisdom's unity sages – from Rumi to Tagore – demonstrate the power of embracing unity.

CHAPTER 15: THE HAVEN OF HIGHER UNDERSTANDING

The haven's gentle breezes carry the whispers of wisdom, guiding us toward our highest potential. Wisdom's gentle guides – from Mary Oliver to Desmond Tutu – demonstrate the power of listening.The haven's tranquil gardens symbolize the serenity of wisdom, where mindfulness and meditation guide us toward inner peace. Wisdom's peaceful sages have long recognized the power of cultivating stillness.

As we explore the haven's vast archives, we discover the treasures of forgotten wisdom, where ancient knowledge awaits rediscovery. The collective wisdom of humanity is a boundless resource, awaiting our exploration.

The haven's luminous stars represent the celestial guidance of wisdom, illuminating our path through life's complexities. By embracing wisdom's timeless principles, we may navigate life's challenges with greater ease.

The haven's majestic waterfalls symbolize the transformative power of wisdom, cascading into the depths of our being. As we embody wisdom's teachings, we become instruments of positive change.

The haven's resplendent bridges represent the connections between diverse perspectives, uniting humanity in a shared understanding. Wisdom's bridge builders have long fostered dialogue and empathy.The haven's gentle morning dew represents the refreshing power of wisdom, nourishing our minds and hearts. Wisdom's gentle guidance revitalizes our spirit.

The haven's vibrant tapestries represent the intricate threads of interconnection, weaving humanity into a single fabric. Wisdom's tapestry weavers have long recognized our shared humanity.

The haven's luminous horizon represents the limitless potential of wisdom, illuminating our path toward higher understanding. Wisdom's illuminating sages guide us toward enlightenment.The haven's peaceful forests symbolize the serenity of wisdom, calming the mind and soothing the soul. Wisdom's peaceful sages have long sought solace in nature's beauty.

17

Chapter 16:The refuge of reflection

As we enter the refuge of reflection, we discover a sanctuary of introspection, where wisdom's timeless principles guide us toward self-awareness and personal growth.The refuge's entrance is flanked by the gates of mindfulness, where meditation and contemplation illuminate our journey. Wisdom's reflective thinkers have long recognized the power of examining our thoughts.

The refuge's serene gardens symbolize the tranquility of reflection, where nature's beauty inspires introspection. Wisdom's peaceful sages cultivate stillness.

The refuge's luminous mirrors represent the clarity of self-awareness, where wisdom's light dispels ignorance. Wisdom's enlightening sages guide us toward higher understanding.

The refuge's majestic library represents the vast repository of human knowledge, where wisdom's timeless principles await discovery. Wisdom's sage scholars have long honored our cultural heritage.

The refuge's peaceful lakes represent the calmness of reflection, where wisdom's tranquil waters soothe the soul. Wisdom's peaceful sages have long sought solace in contemplation.The refuge's vibrant tapestries represent the intricate threads of interconnection, weaving humanity into a single fabric. Wisdom's tapestry weavers recognize our shared humanity.

The refuge's radiant sunlight symbolizes the illumination of insight, where wisdom's light guides us toward personal growth. Wisdom's enlightening sages inspire self-awareness.The refuge's gentle breezes carry the whispers of wisdom, guiding us toward our highest potential. Wisdom's gentle guides listen.As we wander through the refuge's contemplative walkways, we encounter the statues of introspection, where self-awareness and personal growth guide our journey. Wisdom's reflective sages have long recognized the power of quiet contemplation.

The refuge's majestic architecture represents the grandeur of human potential, where wisdom's timeless principles inspire excellence. Wisdom's visionary architects design lives of purpose and meaning.

The refuge's luminous stars represent the celestial guidance of wisdom, illuminating our path through life's complexities. Wisdom's stellar navigators recognize our place within the universe.

The refuge's serene fountains symbolize the soothing power of reflection, where wisdom's tranquil waters calm the mind and

CHAPTER 16:THE REFUGE OF REFLECTION

soothe the soul. Wisdom's peaceful sages cultivate inner peace.

The refuge's vibrant marketplace represents the exchange of ideas, where diverse perspectives enrich our understanding. Wisdom's marketplace sages foster dialogue and empathy.The refuge's gentle morning dew represents the refreshing power of reflection, nourishing our minds and hearts. Wisdom's gentle guides revitalize our spirit.

The refuge's majestic mountains symbolize the towering achievements of human resilience, where wisdom's timeless principles guide us through adversity. Wisdom's resilient leaders inspire hope.

The refuge's luminous horizon represents the limitless potential of reflection, illuminating our path toward personal growth. Wisdom's illuminating sages guide us toward self-awareness.The refuge's peaceful forests symbolize the serenity of contemplation, where wisdom's tranquil waters soothe the soul. Wisdom's peaceful sages cultivate stillness.

18

Chapter 17:The sanctuary of gratitude

As we enter the sanctuary of gratitude, we discover a haven of appreciation, where wisdom's timeless principles guide us toward thankfulness and contentment.

The sanctuary's entrance is flanked by the gates of awareness, where mindfulness and reflection illuminate our journey. Wisdom's grateful sages have long recognized the power of appreciation.

The sanctuary's vibrant gardens symbolize the beauty of gratitude, where nature's splendor inspires thankfulness. Wisdom's peaceful sages cultivate inner peace.

The sanctuary's luminous candles represent the warmth of appreciation, where wisdom's light dispels negativity. Wisdom's enlightening sages guide us toward positivity.

The sanctuary's majestic architecture represents the grandeur of human connection, where wisdom's timeless principles

CHAPTER 17:THE SANCTUARY OF GRATITUDE

inspire empathy. Wisdom's visionary architects design lives of compassion.The sanctuary's serene waterfalls symbolize the soothing power of gratitude, calming the mind and soothing the soul. Wisdom's peaceful sages have long sought solace in appreciation.

The sanctuary's radiant sunlight represents the illumination of thankfulness, where wisdom's light guides us toward contentment. Wisdom's enlightening sages inspire gratitude.

The sanctuary's vibrant tapestries represent the intricate threads of interconnection, weaving humanity into a single fabric. Wisdom's tapestry weavers recognize our shared humanity.

The sanctuary's gentle breezes carry the whispers of wisdom, guiding us toward our highest potential. Wisdom's gentle guides listen.

The sanctuary's peaceful lakes represent the calmness of gratitude, where wisdom's tranquil waters soothe the soul. Wisdom's peaceful sages cultivate inner peace.The sanctuary's majestic mountains symbolize the towering achievements of human resilience, where wisdom's timeless principles guide us through adversity. Wisdom's resilient leaders inspire hope.

The sanctuary's luminous horizon represents the limitless potential of gratitude, illuminating our path toward personal growth. Wisdom's illuminating sages guide us toward self-awareness.

The sanctuary's serene walkways represent the contemplative power of appreciation, where wisdom's tranquil waters calm the mind. Wisdom's peaceful sages cultivate stillness.The sanctuary's beautiful mosaics represent the diverse threads of human experience, woven together in a tapestry of gratitude. Wisdom's master artisans craft lives of purpose.

The sanctuary's soothing melodies symbolize the harmonizing power of appreciation, calming the mind and lifting the spirit. Wisdom's melodic sages inspire joy.

The sanctuary's vibrant flowers represent the blossoming of gratitude, where wisdom's timeless principles nurture growth. Wisdom's gardening sages cultivate inner beauty.

The sanctuary's radiant beams represent the illuminating power of thankfulness, shining light on life's blessings. Wisdom's radiant sages guide us toward positivity.

The sanctuary's peaceful pavilions symbolize the serenity of gratitude, providing refuge from life's storms. Wisdom's peaceful sages offer shelter.The sanctuary's majestic rivers represent the flowing power of appreciation, nourishing our minds and hearts. Wisdom's river guides navigate life's journey.

The sanctuary's luminous stars represent the celestial guidance of gratitude, illuminating our path toward contentment. Wisdom's stellar navigators recognize our place.

The sanctuary's gentle rainfall symbolizes the refreshing power of thankfulness, revitalizing our spirit. Wisdom's gentle guides revitalize.

The sanctuary's vibrant marketplaces represent the exchange of gratitude, where diverse perspectives enrich our understanding. Wisdom's marketplace sages foster empathy.

The sanctuary's serene landscapes symbolize the tranquility of appreciation, calming the mind and soothing the soul. Wisdom's peaceful sages cultivate inner peace.

19

Chapter 18:The horizon of Eternal wisdom

As we reach the horizon of eternal wisdom, we behold the timeless expanse of human understanding, where wisdom's principles guide us toward ultimate truth.The horizon's radiant sunrise symbolizes the dawn of enlightenment, illuminating our path toward higher understanding. Wisdom's enlightening sages guide us.

The horizon's vast landscape represents the boundless potential of human growth, where wisdom's timeless principles inspire excellence. Wisdom's visionary leaders inspire hope.

The horizon's luminous stars represent the celestial guidance of wisdom, shining light on life's journey. Wisdom's stellar navigators recognize our place.The horizon's gentle breeze carries the whispers of wisdom, guiding us toward our highest potential. Wisdom's gentle guides listen.

As we stand at the horizon of eternal wisdom, we realize

CHAPTER 18:THE HORIZON OF ETERNAL WISDOM

the journey is not ending, but evolving. Wisdom's timeless principles remain our guiding light.The horizon's majestic mountains symbolize the towering achievements of human resilience, where wisdom's timeless principles guide us through adversity. Wisdom's resilient leaders inspire hope.

The horizon's serene waters represent the calming power of wisdom, soothing the mind and soothing the soul. Wisdom's peaceful sages cultivate inner peace.

The horizon's vibrant tapestry represents the intricate threads of interconnection, weaving humanity into a single fabric. Wisdom's tapestry weavers recognize our shared humanity.

The horizon's radiant sunlight represents the illuminating power of wisdom, shining light on life's blessings. Wisdom's radiant sages guide us toward positivity.

As we gaze upon the horizon of eternal wisdom, we behold the infinite possibilities of human potential, where wisdom's timeless principles inspire excellence.

Epilogue:

And so, dear traveler, our journey through the world atlas of wisdom comes full circle. May the timeless principles of wisdom guide you on your life's path, illuminating your way toward higher understanding, inner peace, and ultimate truth.As we stand at the horizon of eternal wisdom, we recognize the interconnectedness of all existence. Wisdom's timeless principles weave together the threads of human experience, forming a rich tapestry of understanding.

The horizon's luminous light symbolizes the illumination of wisdom, shining brightly across the expanse of human knowledge. Wisdom's enlightening sages guide us toward the highest truth.

In the distance, we see the silhouette of humanity's collective journey, winding through the landscapes of time. Wisdom's timeless principles have guided us thus far, and will continue to illuminate our path.

As we gaze upon the horizon, we behold the infinite possibilities of human potential. Wisdom's timeless principles inspire excellence, nurturing the seeds of hope and resilience.

The horizon's gentle breeze carries the whispers of wisdom, reminding us of the power of compassion, empathy, and kindness. Wisdom's gentle guides listen, offering guidance.

In this moment, we understand that wisdom is not a destination, but a journey. A journey of self-discovery, growth, and transformation. Wisdom's timeless principles remain our guiding light.

As the sun sets on our journey, we carry with us the wisdom of the ages. May its timeless principles illuminate our path, guiding us toward a brighter future.

The horizon of eternal wisdom beckons, inviting us to continue our journey of discovery. May we walk in wisdom's light, forever guided by its timeless principles.

And so, dear traveler, our journey comes full circle. May the wisdom of the ages remain with you always.

Final Reflection:

As we close this chapter, remember that wisdom's journey is lifelong. May its timeless principles guide you through life's joys and challenges, illuminating your path toward higher understanding, inner peace, and ultimate truth.

20

Epilogue

Epilogue: The Enduring Legacy of Wisdom

As we close this journey through the world atlas of wisdom, we carry with us the timeless principles that have guided humanity throughout the ages. We have traversed the landscapes of human experience, from the depths of introspection to the heights of enlightenment.

The wisdom we've encountered has transformed us, nurturing our minds, hearts, and spirits. We've discovered that wisdom is not merely knowledge, but a way of living – a path that weaves together compassion, empathy, and kindness.

As we embark on the next chapter of our own journey, we remember that wisdom's lessons are eternal. They transcend time and circumstance, speaking directly to our souls.

May the wisdom of the ages continue to guide us:

EPILOGUE

May we walk in humility, recognizing the mystery that surrounds us.

May we cultivate compassion, embracing our shared humanity.

May we seek understanding, honoring diverse perspectives.

May we embody kindness, spreading love and light.

And may we pass on the wisdom we've received, illuminating the path for generations to come.

In the words of the wise:

"Wisdom is not just knowledge; it's the application of knowledge."

"True wisdom comes from experiencing life itself."

"Wisdom's beauty lies in its simplicity."

As we bid farewell to this atlas, remember that wisdom's journey never ends. May its timeless principles remain your guiding light, illuminating your path toward higher understanding, inner peace, and ultimate truth.

Farewell, dear traveler.

May wisdom be your constant companion.

— -

21

Glossary

Glossary: Wisdom Terms and Concepts

A

- Awareness: State of being mindful and present
  - Authenticity: Genuine expression of oneself

C

- Compassion: Empathy and kindness toward self and others
  - Contemplation: Reflective thinking and introspection

E

- Empathy: Understanding and sharing others' feelings
  - Enlightenment: Spiritual awakening and self-realization

H

GLOSSARY

- Humility: Modesty and recognition of limitations
    - Holism: Integrating physical, emotional, and spiritual well-being

I

- Introspection: Self-reflection and examination
    - Insight: Sudden understanding or awareness

K

- Kindness: Gentle and compassionate treatment of others

M

- Mindfulness: Present-moment awareness
    - Meditation: Practice of focused attention and calm

P

- Perseverance: Persistent effort despite challenges
    - Patience: Tolerance and understanding in difficult situations

R

- Resilience: Ability to cope with adversity
    - Reflection: Thoughtful consideration and evaluation

S

- Self-awareness: Understanding one's thoughts, emotions, and behaviors
    - Spirituality: Connection to something greater than oneself

T

- Transformation: Positive change and growth
    - Tolerance: Acceptance and understanding of differences

W

- Wisdom: Insight and understanding guiding life decisions
    - Well-being: Overall physical, emotional, and spiritual health

Y

- Yoga: Integration of body, mind, and spirit through physical postures, breathing, and meditation

Z

- Zen: State of calm and focused awareness

22

Bibliography

Ancient Wisdom Texts

1. Aristotle. (350 BCE). Nicomachean Ethics.
2. Aurelius, M. (167-180 CE). Meditations.
3. Buddha. (500 BCE). The Dhammapada.
4. Lao Tzu. (600 BCE). Tao Te Ching.
5. Plato. (380 BCE). The Republic.

Philosophy and Spirituality

1. Dalai Lama. (1999). Ethics for the New Millennium.
2. Eckhart Tolle. (1997). The Power of Now.
3. Immanuel Kant. (1785). Grounding for the Metaphysics of Morals.
4. Jean-Paul Sartre. (1943). Being and Nothingness.

5. Martin Heidegger. (1927). Being and Time.

Psychology and Self-Development

1. Carl Jung. (1921). Psychological Types.
2. Daniel Goleman. (1995). Emotional Intelligence.
3. Maslow, A. H. (1943). A Theory of Human Motivation.
4. Sigmund Freud. (1923). The Ego and the Id.
5. Viktor Frankl. (1946). Man's Search for Meaning.

Modern Wisdom Literature

1. Brené Brown. (2010). The Gifts of Imperfection.
2. Charles Tart. (1986). Waking Up.
3. Eckhart Tolle. (2005). A New Earth.
4. Ram Dass. (1971). Be Here Now.
5. Thich Nhat Hanh. (1975). The Miracle of Mindfulness.

Academic Journals

1. Journal of Positive Psychology.
2. Journal of Philosophy and Psychology.
3. Mindfulness: A Practical Approach to Finding Peace.

23

Acknowledgment

Acknowledgments

I am deeply grateful to the following individuals and organizations for their contributions to this book:

Special Mentions

- My family, for their unwavering support and patience.
    - My editor, [Taylor smoke], for expert guidance.
    - My research assistant, [Department of Psychology], for diligent efforts.

Wisdom Keepers

- Dalai Lama, for inspiring compassion and wisdom.
    - Eckhart Tolle, for illuminating the power of now.
    - Brené Brown, for sharing the gifts of imperfection.

Colleagues and Friends

- [Matemane Thabiso,Mashaba Msindeni,for valuable feedback and encouragement.
    - [Prince Phaahle], for insightful discussions.

Organizations

- The Wisdom Project, for sharing resources.
    - TED Talks, for spreading wisdom.

Publishing Team

- [Tshireparadise], for believing in this project.
    - [Prince Phaahle], for creating a beautiful design.

Readers

- Thank you for embracing this wisdom journey.

May this book honor the wisdom of those who have come before us and inspire future generations.

Sincerely,

[Tshireletso Prince Phaahle]

24

Final thoughts

May the wisdom shared in these pages guide you on your journey toward greater understanding, compassion, and inner peace.

Remember, wisdom is a lifelong pursuit, and its journey is just as important as its destination.

Embrace the beauty of uncertainty, and may curiosity be your constant companion.

May you cultivate mindfulness, empathy, and kindness in all aspects of your life.

And when the winds of adversity blow, may you find strength in the timeless principles of wisdom.

As you close this book, remember that the true wisdom lies within yourself.

Trust your heart, listen to your soul, and may your path be

illuminated by the light of wisdom.

Farewell, dear reader. May our paths cross again.

Sincerely,
   [Tshireletso Prince Phaahle]

25

Conclusion

As we conclude our journey through the world atlas of wisdom, we carry with us a profound understanding of the timeless principles that guide human flourishing.

We've explored the landscapes of introspection, compassion, and kindness, and discovered the transformative power of wisdom.

Through the wisdom of sages, philosophers, and everyday heroes, we've learned to:

- Cultivate mindfulness and presence
    - Embody compassion and empathy
    - Navigate life's challenges with resilience
    - Seek wisdom in uncertainty

May the wisdom shared in these pages:

- Illuminate your path

- Guide your decisions
- Nurture your spirit
- Inspire your highest potential

As you integrate these principles into your life, remember:

Wisdom is a journey, not a destination.
 Its power lies in its application.
 Its beauty lies in its simplicity.

May you walk in wisdom's light, and may your life be a testament to its transformative power.

26

Endnotes

Chapter 1: The Journey Begins

1. "Wisdom is the supreme part of happiness." - Aristotle, Nicomachean Ethics (Book 6, Chapter 13)
2. "The unexamined life is not worth living." - Socrates, The Apology (38a)

Chapter 2: Introspection

1. "Know thyself." - Ancient Greek inscription at Delphi
2. "Self-awareness is the ability to take an honest look at yourself." - Daniel Goleman, Emotional Intelligence

Chapter 3: Compassion

1. "Compassion is the heart of wisdom." - Dalai Lama, Ethics for the New Millennium
2. "Empathy is the bridge that connects us." - Brené Brown, The Gifts of Imperfection

Chapter 4: Resilience

1. "Fall seven times, stand up eight." - Japanese proverb
2. "Resilience is the capacity to recover quickly from difficulties." - American Psychological Association

Chapter 5: Wisdom in Action

1. "Wisdom is the application of knowledge." - Unknown
2. "The wise person is the one who puts into practice what they know." - Aristotle, Nicomachean Ethics (Book 6, Chapter 13)

Additional Endnotes:

1. "The power of now." - Eckhart Tolle, The Power of Now
2. "Mindfulness is the practice of being present." - Jon Kabat-Zinn, Wherever You Go, There You Are!!

27

Author information

About the Author

[Tshireletso Prince Phaahle]

Am curious and passionate individual dedicated to exploring the complexities of human experience. With a deep fascination for wisdom, philosophy, and personal growth, I've spent years studying the intersection of psychology, spirituality, and culture.

As a writer, I aim to share insightful and thought-provoking content that inspires others to embark on their own journey of self-discovery and transformation. My work is informed by timeless wisdom, cutting-edge research, and everyday moments of beauty and resilience.

When not writing, you can find me on WhatsApp (0662312576)I'm grateful for the opportunity to connect with like-minded individuals and share my perspectives with the world.

Stay connected with me on [https://35be9cc869.authorwebsites.bookbub.com/]

— -

  Contact Information

- Email: [phaahleprince9@icloud.com
  - Website: [https://35be9cc869.authorwebsites.bookbub.com/]

Author's Note

I am passionate about exploring the human experience and sharing wisdom to inspire personal growth. This book represents my journey to understand the complexities of life and the transformative power of wisdom.

Stay Connected

Join my newsletter for updates, insights, and exclusive content: [https://www.linkedin.com/redir/redirect?url=https%3A%2F%2F35be9cc869%2Eauthorwebsites%2Ebookbub%2Ecom%2F&urlhash=p6iJ&trk=contact-info]

online store[https://payhip.com/Tshireparadisebookclub]

Acknowledgments

Grateful thanks to my family, friends, and colleagues for their support and encouragement.

Copyright Information

©2024 Tshireletso Prince Phaahle. All rights reserved.